CRITICAL THINKER'S GUIDE *to* THE ENVIRONMENT

CRITICAL THINKER'S GUIDE *to* THE ENVIRONMENT

Martin M. Kaufman
The University of Michigan–Flint

John Wiley & Sons, Inc.
New York • Chichester • Brisbane • Toronto • Singapore

ISBN 0-471-32231-8

Printed in the United States of America

10 9 8 7 6 5 4 3 2 1

To my brothers Alan and Bob

PREFACE

When *A Critical Thinker's Guide to the Environment* (CTG) was being conceived it came down to this: What does being critical mean? It was felt that being critical within the context of environmental issues and problems meant being objective and comprehensive, while using scientific reasoning or the scientific method. The CTG asks students not only to accept that human activity and the environment are interrelated, it challenges them to observe and to think about the nature of those interrelations.

We frequently make environmental decisions by combining our current knowledge only with what we have seen. Too often, we neglect the other tools of scientific investigation, such as classification, the use of theory, analysis, and research. This book has been designed to introduce students to these tools from a geographic perspective, and to provide examples of their concurrent application.

Many people helped with the preparation of this book. Initial thanks go to William M. Marsh and John M. Grossa Jr., who wrote *Environmental Geography: Science, Land Use, and Earth Systems*, and supported the creation of a non-traditional companion guide. William M. Marsh provided excellent comments on the manuscript and several original pieces of artwork. The innovative "food chain geography" application in Chapter 6 was his idea. Nina Marsh also provided many useful suggestions on format and content. In addition, the manuscript reviewers, Jack Kinworthy of Concordia College, Nebraska, Jill Boberg of the University of Maryland, and Scott Carlin of Long Island University offered many helpful suggestions on format and content.

I would also like to thank Eric Stano and Jennifer Bruer, editors at Wiley, and Bethany Brooks, editorial assistant, for their helpful suggestions and support throughout the project.

About the Author

Martin M. Kaufman is Assistant Professor of Earth and Resource Science at the University of Michigan-Flint. He has a Ph.D in Environmental Planning from the University of Michigan in Ann Arbor. Currently, his research is focused on stormwater management and watershed restoration efforts in urbanized watersheds. His professionial experience includes extensive environmental analysis, geographic information systems development, and participation in large-scale watershed restoration efforts.

INTRODUCTION (for instructors)

A primary goal is to make *A Critical Thinker's Guide to the Environment* (CTG) easy for the instructor and student to use. The CTG consists of 16 chapters corresponding to the chapters of Marsh and Grossa's *Environmental Geography: Science, Land Use, and Earth Systems,* with a format designed to facilitate cross-reference between the two books.. For each chapter subsection in *Environmental Geography,* there are similarly numbered review sections in the CTG used to illustrate and apply concepts. A minor exception is the chapter summary in the CTG used in place of the Introduction section in *Environmental Geography.*

As part of *Environmental Geography,* the CTG will draw largely from the concepts contained there. Additional skills, such as map reading, spatial analysis, and matrix use are introduced within the context of the review sections. The approach taken is interdisciplinary; the perspective geographic.

Key concepts are highlighted at the beginning of each subsection; and these concepts also appear in the margin where they are discussed in the body of the text. Critical thinking questions are posed throughout, and these are underlined to help them stand out. Vignettes (some from outside authors) are used in most chapters to illustrate additional concepts. After each vignette is a "Follow-up" section denoted by a magnifying glass in the margin. The consolidations of the lessons from the vignettes are presented here.

After the review sections, there is an "Applications/Discussion" section numerically keyed to the review sections (since no applications exist for the summary section of each chapter--which is section "1"--the question numbers begin with "2"). Items in this section include fieldwork exercises, class discussion questions, and a mixture of long and short answer questions covering the concepts of each review section. The use of a variety of question formats is designed to allow the instructor more flexibility in lecture and assignment preparation.

So, this is not your traditional study guide. The emphasis is on concepts. There are no long lists of example test questions or terms for review. The CTG is designed to allow students to learn from their observations of environmental processes after the final exams are taken.

The basic approach to learning taken by the CTG is shown below:

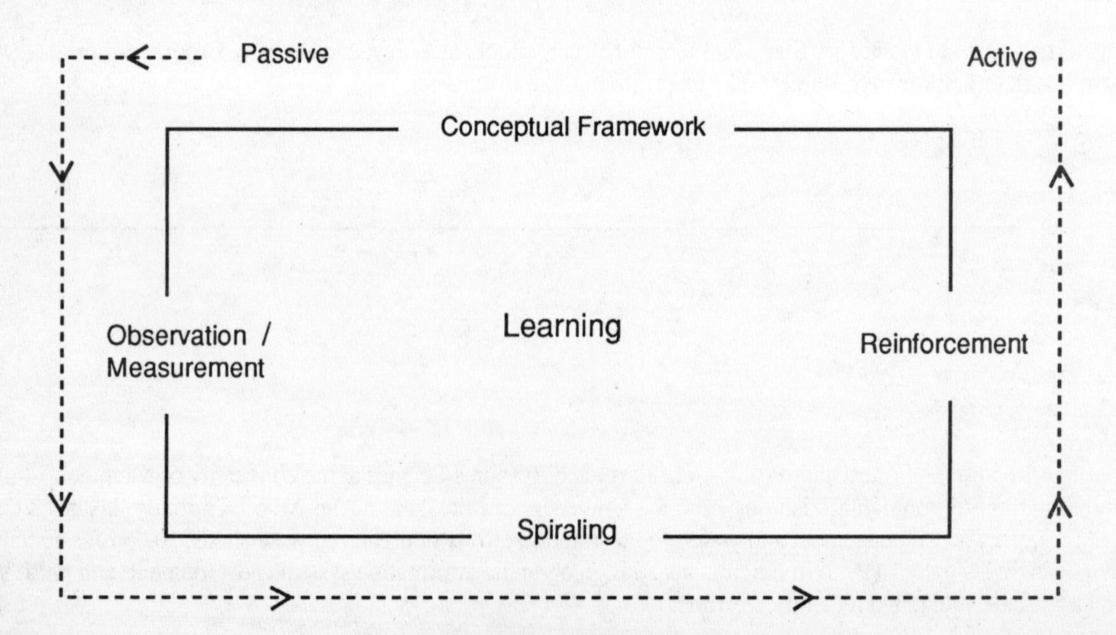

The CTG provides students with conceptual frameworks, then allows them to apply these through observation and measurement to environmental issues and problems. Spiraling involves introducing a concept multiple times, first at a general level, then later in a more detailed manner. This technique has been used successfully to teach science and higher math. Reinforcement is accomplished within and between chapters by applying similar concepts to different situations.

INTRODUCTION (for students)

Most of us often rely solely on the passive use of our sensory functions, especially sight, to make decisions concerning how the environment works and how it is managed. Quite often, the result is an incorrect response. The indiscriminate draining of wetlands—once considered visual nuisances and health threats, is an example.

The situation today is even more critical because of the prevalence of complex technological mediators between environment and observer. Increasingly, we have little opportunity to engage in the world of real environmental problems because: 1) our lives are being played out in abstract "virtual" realities; and, 2) we are not empowered with the powers of observation to think critically about the world around us because we have fewer opportunities to engage the environment in a give and take manner.

Most of us have passive environmental experiences. We mow the lawn, ski, fish, etc. These activities, for the most part, do not allow a critical understanding of the environment, nor are most of us today thinking much about them as environmental activities in the first place.

A Critical Thinkers Guide to the Environment (CTG) will help students in the environmental sciences develop insight into the role of humans in ecosystems not only from a passive perspective, but actively through participation in the observation and measurement of natural and social processes.

CONTENTS

1

THE CRITICAL CHALLENGE

Review

1.1 Summary

There are three general components to the environmental dilemma. The first component is the growing human population, which over time has produced greater impacts on the earth's natural systems. As technology and population have increased, so have the amount and extent of pollution and disturbance, creating complex systems of negative impacts.

The second and third components relate to the characteristics of earth's basic natural resources—air, water, land, and their associated minerals. These resources are unevenly distributed, and many, such as land itself, are finite. The predominance of salt over fresh water, narrow climatic ranges for prime soil development, and inadequate precipitation are some of the objective limitations on potential biological productivity and the range of human habitat.

Thus the critical challenge facing humans is how to manage the geographic expansion and growing numbers of their species, while maintaining a sustainable balance with the environment.

1.2 The Global Context and Setting

Concepts: location, zones/regions, scale

location

Our concern with the environment at the broadest level is the entire earth—its air, land, water, and people. In order to answer basic questions about location, such as: where do people live, and how certain minerals are distributed, we have constructed global reference systems. Latitude and longitude is one global reference system allowing us to have a common language for communicating location.

zones/regions

Sometimes, a broader locational framework is more meaningful. Zones, or regions, help to organize large amounts of spatial information. Each type of region has characteristics, or attributes, making it unique. Examples of regions include those of area, such as the hemispheres, and regions of climate.

Suppose you were taking a trip. You called some friends on the phone to tell them where you were going, but you could not pronounce the name of the destination. What could you tell them that would allow them to figure it out for themselves, and allow you to talk about what you plan to do when you get there?

You say: “It's just north of the equator, in the eastern hemisphere.”
“Great,” your friends reply, “that's a quarter of the world. Ay, where are you going?”
“The climate is real warm there, and it rains frequently in the winter.”
“Are you going to India?”
“Yes. I'm flying into Tiruchchirappalli.”

This short conversation demonstrates how four large hemispheres, or zones, were created from the equator and Prime Meridian. Additional information, such as climate (the monsoon rains), helped to identify the location.

In general, physical barriers, physical processes, and the need for specific types of information facilitate regional formation. In the case of barriers, desert regions are often found in areas on the lee side of mountain ranges, since the air is warmed and dried as it descends. Certain physical processes, such as the movement at the margins of the large tectonic plates in the earth's crust, create seismic zones. And, to help scientists study climate, different bioclimatic regions are created based upon their temperature, moisture pattern, and biotic composition. Regions thus vary considerably in extent, or scale.

scale

Essentially, there are three meanings of scale. First, scale is the spatial extent of a study, often referred to as geographic scale, or the scale of observation. A study of global wind patterns would be called a large scale study. Second, cartographic scale is the ratio of map distance to earth distance (Fig. 1.1). The third usage of scale refers to the operational scale, or extent at which a process operates. For example, while stream bank erosion is localized to a specific stream valley, the location of the ozone hole over Antarctica results from global wind patterns.

Scale is one of the central "perspectives" of geography. The examples above demonstrate how multiple disciplines can be viewed from a scale perspective.

1.3 Population and Earth's Carrying Capacity

Concepts: rates, population growth, carrying capacity, sustainability

When looking for descriptions of human characteristics, we often consult nature. Did you ever hear somebody say: “He/she's growing like a weed?” This description communicates to us an image of something that occurs over a span of weeks or months; something we don't notice all of the time. Often the answer to the question above is: “I see what you mean”—but that person never really saw the person grow.

rates

The concept described above is a rate. A rate is a change in the quantity of something in a given amount of time. How would you describe the recent growth of the earth's population? Is global population growing like a weed? When will the population reach full height? It might be tough to describe population growth like that, so we use other ways to show the rate of population growth, such as graphs.

Figure 1.2 shows how population tends to grow—which by the way, was borrowed from laboratory studies of how bacteria grow.

population growth

The three labeled components of the graph parallel the historical phases of human population growth. The growth lag phase corresponds to the 100,000 years it took to grow a population of 2 billion people. The very rapid, or exponential increase represents the 46 years to double the population from 2 to 4 billion, and the leveling off portion

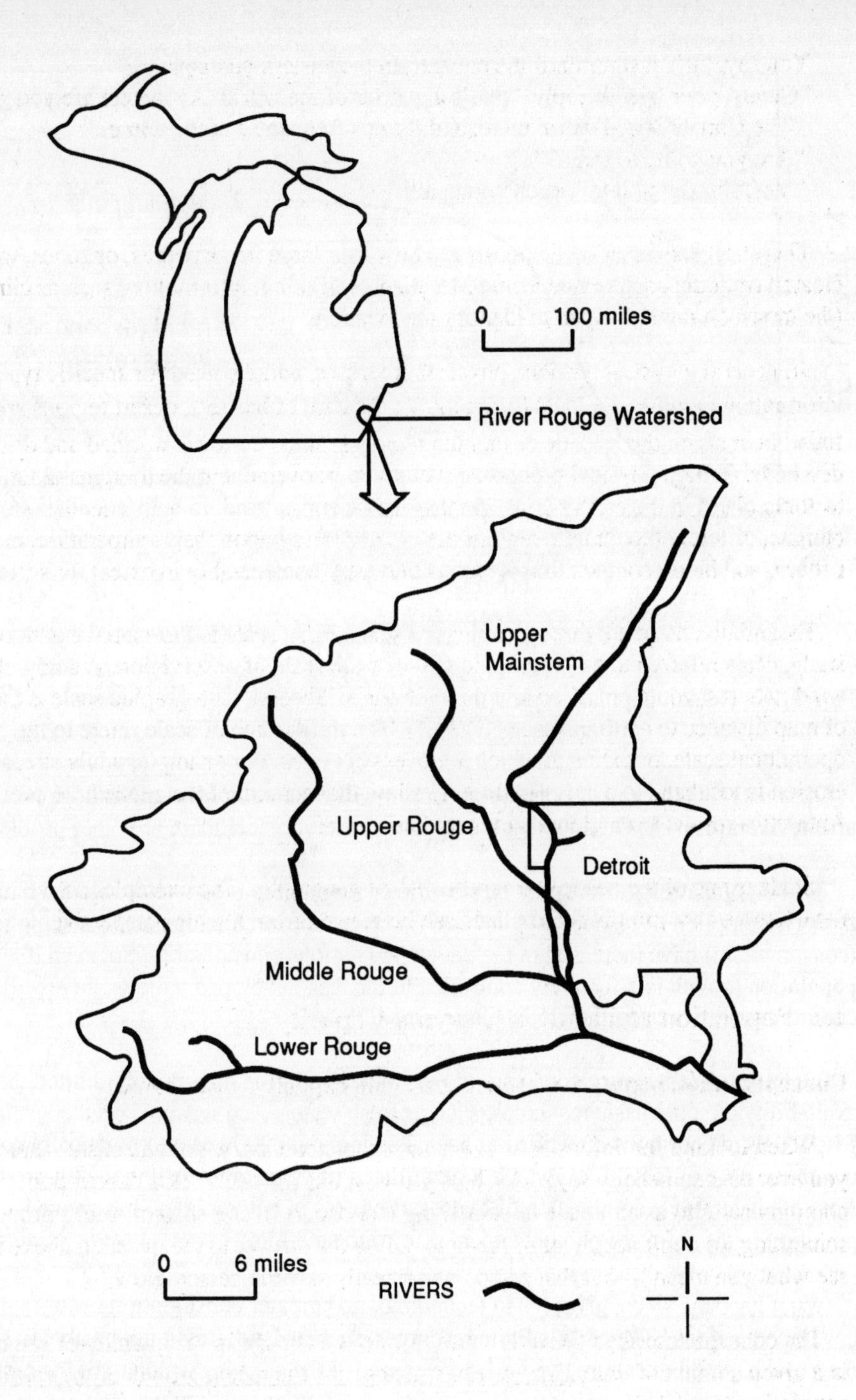

Figure 1.1 Cartographic map scale

conforms to the growth pattern in the more developed countries only.

This distribution of population growth is a major concern. The developed nations have experienced a deceleration of their population growth, yet they represent only 1/5 of world's total population. Thus, the leveling off phase has not yet occurred with respect to the global population growth rate.

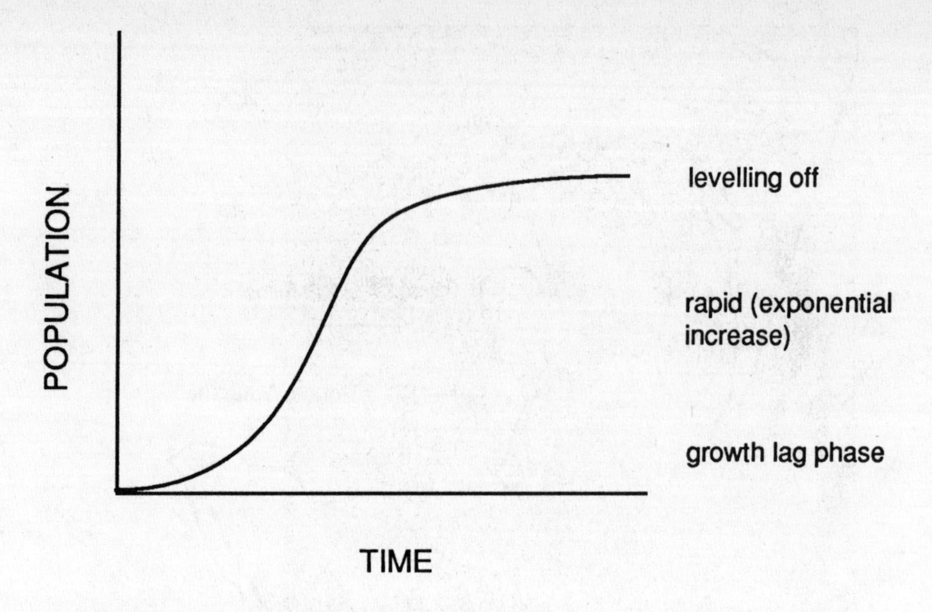

Figure 1.2 Human population growth

Optimists feel the global population pattern will eventually level off and we will be able to avert disaster. On the other hand, many observers feel there are other problems.

While population growth may be slowing or stagnant in the more developed nations, their consumption of resources is increasing. Over time, technology and consumption (consumerism) have increased in the developed nations considerably. So, even if population growth is effectively controlled in the less-developed nations, there will be continuing pressures on the global environment. Why?

We have not taken very good care of our home. Pollution has actually limited the availability of some basic resources, such as fresh water. And we have been consuming fossil fuels in enormous quantities. It has taken less than 200 years to practically deplete the entire store of fossil fuels which took millions of years to form; and the by-products from the incomplete combustion of hydrocarbons (as in automobiles) has created pollution in the forms of photochemical smog and acid rain.

carrying capacity

What happens when advances in technology and greater consumption occur in the less developed nations? The earth's finite resources, such as land, might not be able to sustain the growth. At this juncture, we will have exceeded the carrying capacity of the earth.

In order to avoid disaster, we need to reach some kind of sustainable balance. What kind of balance?

sustainability

The achievement of sustainability rests upon the ability to balance human needs and available resources.

Have you ever been on a treadmill? (Fig. 1.3). This is a good way to imagine global environmental sustainability. In order for us to have a sustainable environment, we need to achieve a balance the way a person does on a treadmill. If you wanted to do a 20 minute workout, you wouldn't set the speed so high it caused you to lose your balance

Figure 1.3 Person on a treadmill

and fall off, or to become exhausted in less than 20 minutes.

Since we need our environment to support its life longer than 20 minutes, achieving global balance is a bigger task. Here, balance is achieved when humans use natural resources at a rate and in a manner which does not prevent or hinder the reuse of those same resources indefinitely into the future. One way to prolong the environment's endurance is by minimizing human impacts—disturbance and pollution.

1.4 The Four Elements of the Dilemma

Concepts: sustainable land use systems, interrelationships

Everything we do has an impact on the environment. Our breathing expels carbon dioxide into the atmosphere. At larger scales humans have displaced and replaced entire natural habitats. Agriculture has been a major disturbance on the original landscape, as forests once covering thousands of square miles have been replaced by farms.

Could you logically argue for a cessation of agriculture? The key question for sustainability is how do we limit excessive disturbances, and what are acceptable disturbances?

sustainable land use systems

One line of thought here is to accept the large-scale land-use systems (e.g., agriculture, mining, rangelands, forestry, urban areas) as requirements for continued human occupancy of the planet. The sustenance of human species and other species would then depend upon the ability to employ sustainable practices within and between these land use systems. This idea is catching on.

There is a growing awareness of sustainable agriculture, and several recent books have addressed sustainable cities. We can look at population, technology, and consumerism as a few of the interrelated components which need to be managed in order to achieve sustainability. The big question is: How do you "manage" population, technology, and consumerism?

interrelationships

As science tried to answer this question, it became apparent that the problems were based on the types of relationships which exist between of our human (social) and environmental (natural) systems. It also became apparent that all living organisms and their environment were interrelated.

No man or woman is an island.

1.5 Scope of Environmental Geography

Concepts: interdisciplinary approach, observation

interdisciplinary approach

It is the goal of environmental geography to investigate and understand the relationships between human and natural systems. There is a complex set of physical, biological, and social interactions, and this requires an interdisciplinary approach to environmental analysis. For instance, it would be impossible to fully understand acid rain without knowing where the pollution originated (geography), what the prevailing wind patterns were (climatology), the type of tree species affected (botany), and the conditions of the soil receiving the precipitation (soil science-pedolgy, chemistry).

observation

Though it sounds simple, observation is an effective technique for preventing single-minded approaches. Reliance upon one approach or field can lead to "tunnel vision." The symptoms result from a failure to account for the full range of processes operating to produce an event. In addition, preconceived expectations can lead to specialized searches that miss the clues to understanding (Vig. 1.1).

Vignette 1.1 "There's Nothing Out There," by William Least Heat Moon

Straight as a chief's countenance the road lay ahead, curves so long and gradual as to be imperceptible except on the map. For nearly a hundred miles due west of Eldorado, not a single town. It was the Texas some people see as barren waste when they cross it, the part they later describe at the motel bar as "nothing." They say, "There's nothing out there."

Driving through the miles of nothing, I decided to test the hypothesis and stopped somewhere in western Crockett County on the top of a broad mesa, just off Texas 29. At a distance, the land looked so rocky and dry, a religious man could believe that the First Hand never got around to the creation in here. Still, somebody had decided to string barbed wire around it.

No plant grew higher than my head. For a while, I heard only miles of wind against the Ghost (the author's name for his van), but after the ringing in my ears stopped, I heard myself breathing, then a bird note, an answering call, another kind of birdsong, and another: mockingbird, mourning dove, an enigma. I heard the high zizz of flies the color of gray flannel and the deep buzz of a blue bumblebee. I made a list of nothing in particular.

1. mockingbird 2. mourning dove 3. enigma bird (heard not saw) 4. gray flies 5. blue bumblebee 6. two circling buzzards (not yet, boys) 7. orange ants 8. black ants 9. orange-black ants (what's been going on?) 10. three species of spiders 11. opossum skull 12. jackrabbit (chewed on cactus) 13. deer (left scat) 14. coyote (left tracks) 15. small rodent (den full of seed hulls under rock) 16. snake (skin hooked on cactus spine) 17. prickly pear cactus (yellow blossoms) 18. hedgehog cactus (orange blossoms) 19. barrel cactus (red blossoms) 20. devil's pincushion (no blossoms) 21. catclaw (no better name) 22. two species of grass (neither green, both alive) 23. yellow flowers (blossoms smaller than peppercorns) 24. sage (indicates alkali-free soil) 25. mesquite (three-foot plant with eighty-foot roots to reach water that fell as rain two thousand years ago) 26. greasewood (oh, yes) 27. joint fir (steeped stems make Brigham Young tea) 28. earth 29. sky 30. wind (always).

That was all the nothing I could identify then, but had I waited until dark when the desert really comes to life, I could have done better. To say nothing is out here is incorrect; to say the desert is stingy with everything except space and light, stone and earth is closer to the truth.

Observation not only makes for a more interesting trip, it is an important part of the scientific method. Great discoveries—such as penicillin and oxygen, have resulted from simply noticing and recording the unexpected, rather than discarding it as "noise" or "interference." Who could practice astronomy without observation?

1.6 The Human Values Factor

Concept: values

values

How we "see" something depends on many factors. There are physical aspects of observation, such as using binoculars, which might allow us a better view. We also interpret what we have seen or heard, and it is here where significant differences arise between people and governments on what they have seen, and on how they should respond. How often have we seen in the news where some people refuse to leave their homes despite the imminent danger from an approaching hurricane?

How we respond to environmental events, and to the environment in general reflects our values. For instance, what do we value in our rivers?

Apparently we value the economic potential of rivers, but our values do not necessarily reflect good scientific judgement. Over 70 percent of the major rivers in the northern third of the world (U.S., Canada, Northern Europe, and Northern Asia) have been dammed for hydropower and flood control. In many rivers, this has produced major environmental disturbances, such as habitat fragmentation, interruption of migratory routes, breeding and nesting areas, increased erosion, and water temperature fluctuations. Many of these problems may be irreversible.

Large-scale land development provides another example of the conflict between the values of short-term gain and long-term sustainability. What was, and is, valued over the human rights of the native peoples of North America, and elsewhere?

1.7 The Geographical Factor

Concepts: geography's unique approach to the environment, spatial analysis

The geographical aspect of the environmental dilemma discussed in section 1.1 confers a special role to geography within the environmental sciences

geography's unique approach to the environment

Geography is interdisciplinary by nature because it approaches problems and issues with a perspective, rather than topical approach. This perspective centers on the distributional, or spatial aspect of earth phenomena, that is, how they are laid out over the earth's surface, no matter what those phenomena may be. Thus geography examines land use, population, water, and soil with equal ease (no matter what other fields claim them) because they are all earth surface phenomena.

spatial analysis

The geographical approach is especially advantageous in environmental studies because many different phenomena, such as land use, air pollution, and acid rain, which are spatially related, can only be understood when their distributions are examined together.

Table 1-1 shows the major areas of geographic inquiry and some of the disciplines which have contributed to these studies.

1.8 Human Use of the Earth

Concepts: More Developed Coutries, Less Developed Countries, environmental hazards

More Developed Countries/Less Developed Countries

Frequently, the level of technology attained by agriculture is used as a measure of national economic development. Less developed countries (LDCs) are characterized by subsistence agriculture—production of food for the family working the land, not the market. More developed countries (MDCs) tend to contain relatively large agricultural areas devoted to commercial production for a market.

TABLE 1-1 The Four Traditions of Geography and Some Contributing Disciplines

Tradition	Contributing Disciplines
1. Regional Studies (from the Greeks)	anthropology, archaeology
2. Physical Geography	geology, chemistry, hydrology, physics, climatology
3. Human/Environmental Interaction	biology, ecology, sociology, epidemiology, economics, political science
4. Spatial Analysis (order, density, pattern, distance, direction, shape, scale, cartography)	art, architecture, mathematics
adapted from Pattison, (1964).	

There are numerous other differences between LDCs and MDCs. We have already mentioned the relative rates of population growth, but health care indicators and literacy rates are other important distinctions.

When making distinctions like this, we tend to focus on the most striking or shocking events, that is, those that are the most obvious, or "newsworthy." Sometimes this relegates important pervasive events to the back burner, and prevents us from approaching the problem comprehensively.

Let's consider health care again. There is currently a major effort to control health care costs in the United States. Many people have begun to advocate preventative medicine and holistic approaches. The costs incurred from treating the effects of smoking and obesity have been cited as major drains on health care resources. This is a good start, but it omits another major use of resources. From the ensuing debate over health care, can you name the single reason why most people miss work, and what costs the most money in terms of the loss of productivity? It is back injuries.

environmental hazards

A similar situation exists with respect to environmental hazards. Environmental hazards occur when humans occupy a place subject to a destructive natural event. Thunderstorms kill more people every year than earthquakes, but we tend to take thunderstorms for granted, and thus avoid taking precautions. Applying a medical analogy to another hazard, erosion is the "silent killer" in the environment. Entire nations such as the Dominican Republic and Nepal are threatened with the loss of their entire forest reserves from soil erosion.

1.9 Prospects for Sustainable Land Use

Concept: growth vs. sustainability

As population grows, and resource consumption increases, the magnitudes and frequencies of environmental disturbances and pollution also increase. This has prompted many people to question whether sustainable land use systems and population/economic growth are compatible.

growth vs. sustainability

This apparent contradiction between growth and sustainability has led many to advocate a return "back to nature." Technology has been abused, and we need to go back to practices that require low inputs of energy. But there are other elements besides technology to consider, such as the correlation existing in many peoples' minds between higher technology and a higher standard of living. It has to be considered whether these practices are socially acceptable, and how to make them acceptable.

Another way of looking at the debate between growth and sustainability is to view it as the difference between those who want to expand production to meet rising demand, and those who would implement measures and controls to manage demand. In the western United States, great expense and environmental damage have occurred simultaneously as large scale farming diverts massive quantities of water from the Colorado River for irrigation. The areas producing food have increased, but the Colorado River has been so depleted of water by irrigation its flow does not even make it to its final destination—the Gulf of California.

In this region, some of the long term environmental impacts are beginning to offset the gains from increased crop production. Most notably, the salinity of the water and soil have increased from over-irrigation. Evaporation of water near the surface in arid regions leaves behind salts and has residues of fertilizers and pesticides which are transported into the river. Over time, this also decreases the fertility of the soil, an interesting paradox, since fertilization is supposed to increase yield. You can simulate this effect by evaporating a jar of water in your home, and observing the residue left on the sides and bottom.

As a result of these environmental impacts, the original benefits are being eaten up by the costs of land reclamation, and the maintenance of a desalination plant near the Mexican border. Under an agreement with Mexico, the United States is obligated not to exceed certain levels of salinity (1000 parts per million--ppm) in the Colorado River as it enters Mexico.* The salt content in the river was often over 1000 ppm, so a desalination plant was built at huge expense ($256 million).**

These events raise several questions; were these diversions necessary to sustain (feed) our nation; what other means were available for providing the same amount of food; and, are we overproducing food in the first place?

Our prospects may depend on our perspective. If hindsight is always 20/20, then we should concentrate on improving the vision of our foresight.

---

* As a reference, the salinity of ocean water is 3500 parts per million (ppm).

** The plant was operational from May 1992 to May 1993, but was taken off-line until a congressional review of its funding was completed.

1.10 Summary: Limits to Growth

Concepts: limits to growth, scale

In its most simple representation, there are three elements to the environmental dilemma (Fig. 1.4).

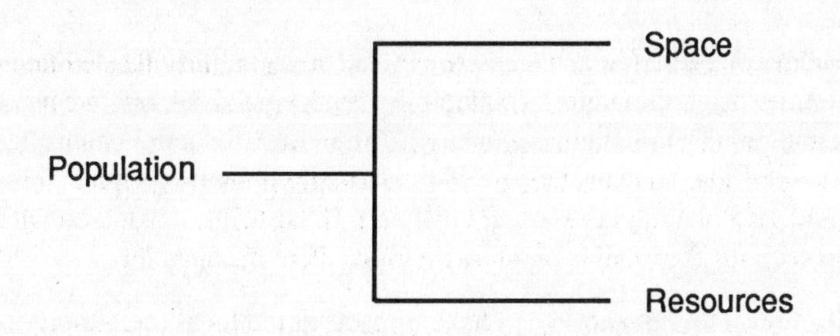

Figure 1.4 Elements of the environmental dilemma

What are the key aspects of population, space, and resources operating to limit growth at the global scale? How do these factors change at smaller scales of observation, such as within your country, or city?

limits to growth, scale

NOTE: In the applications/discussion section of this, and all chapters which follow, the question numbering begins with a "2" because there are no corresponding questions for the initial summary sections.

Applications/Discussion

1.2) How are predictions of population growth made and used? Examine a local city or regional plan and, a) identify the methods used to predict population; b) list two ways the predictions are, or can be used.

1.3) Take a drive to "nowhere" and record what is there. What is your conclusion about what is there; e.g., is the hypothesis of "nothing" correct?

1.4) Give an example of a human impact on the environment that crosses multiple media, that is, one that disturbs air, land, and water resources simultaneously.

1.5) What is harder to change, human values or the physical environment? For instance, is it harder to get people to recycle, or to build larger landfills? How does observation affect the decision-making process with respect to the issues of recycling and landfills?

1.6) Observe a nearby lake or stream. Record the spatial characteristics of the shore or channel. See example below:

Spatial Characteristics	Observation
size	20 feet wide...
depth	
elevation to adjacent land	
slope of adjacent land	
shape	
etc.	

1.7) Describe how a thunderstorm can have impacts at the local (neighborhood), regional (city/county), and national scales.

1.8) Identify another environmental problem besides soil erosion which is pervasive, but on the "back burner."

1.9) Why is reducing energy consumption an important part of achieving sustainable land use systems?

1.10) Why might a city like Seattle, which averages over 60 inches of precipitation/year, experience shortages of drinking water?

Recommended Reading

McDowell, J., and R. Woodbury. (1991). "A Fight Over Liquid Gold," *Time*, 22 July.

Moon, William Least Heat. (1982). *Blue Highways: A Journey into America*. Ballantine, New York.

Reisner, Marc. (1986). *Cadillac Desert: the American West and its Disappearing Water*. Viking, New York.

Sources of Additional Information

INTERNET: Population menu on EnviroGopher

Centre for Resources and Environmental Studies, Australian National University, GPO Box 4, Canberra ACT 2601, Australia
e-mail: geek@cresl.anu.edu.au

Population Institute, 100 Maryland Avenue NE, Suite 207, Washington, D.C. 20002
(202) 544-3300

United Nations Population Fund, 220 E. 42nd Street, New York, NY. 10017

References

Burton, I., R.W. Kates, and G.F. White. (1978). T*he Environment as Hazard.* Oxford University Press, New York, NY.

Dynesius, M., and C. Nilsson. (1994). "Fragmentation and Flow Regulation of River Systems in the Northern Third of the World," *Science*, (266), 753-762.

Lam, N., and D.A. Quattrocchi. (1991). "On the Issues of Scale, Resolution, and Fractal Analysis in the Mapping Sciences," *Professional Geographer*, (44), 88-98.

Pattison, W.D. (1964). "The Four Traditions of Geography," *Journal of Geography* (63), 211-216.

Shryock, H.S. (1976). *The Methods and Materials of Demography*. Academic Press, New York, NY.

2

HUMANS AND EARTH: THE GREAT EXPERIMENT

Review

2.1 Summary

One remarkable aspect of the human experience on earth has been our ability to adapt to different physical environments. Since our species evolved, humans have spread to all of the continents and most islands of the world.

Due to climatic and physical barriers, humans have not been as successful in some areas. These are the "frontier environments" where the availability of moisture, extreme temperatures, difficult terrain, and the incidence of diseases have acted to limit human settlement. Based on these bioclimatic factors, and their current level of economic development and settlement, five types of frontier lands are identified: the wet tropics, the dry lands, the cold lands, the high mountains, and the continental shelves.

As technology has been incorporated into the cultural fabric of nations, the ability to overcome and adapt to these harsh environments has increased, often with damaging results. Attempts to economically exploit wilderness environments by technologically advanced nations has led to severe environmental disturbances and degradation of indigenous societies.

2.2 Human Dispersal and Adaptation

Concepts: diffusion, innovation, adaptation

diffusion

Human diffusion is the spread of people, ideas, and innovations from their sources to other areas. Today, the human population is dispersed widely across the earth, and highly concentrated in a relatively small number of regions. Only Antarctica and the interior of Greenland are uninhabited, yet roughly one-half of the world's population lives concentrated within 5 percent of the remaining available land (primarily in East and South Asia, Western Europe, and Eastern North America).

How did human diffusion occur?

First off, everybody didn't live in the same place, and we made things up as we went along. When food ran out in the neighborhood, people moved on to look for more. In response to cold temperatures, we discovered fire and created clothing. Humans were able to develop a culture, which enabled us to preserve ideas and build on our earliest experiences with the natural landscape.

Thus, the early cultural development of humans was spurred by environmental necessity. In this task we were helped along by four capabilities: observation, exploration, experimentation, and adaptation. The ability to perform these functions explains the success we have had preserving and expanding the range of our species.

Figure 2.1 is a conceptual diagram of what was happening to early people as they struggled to survive.

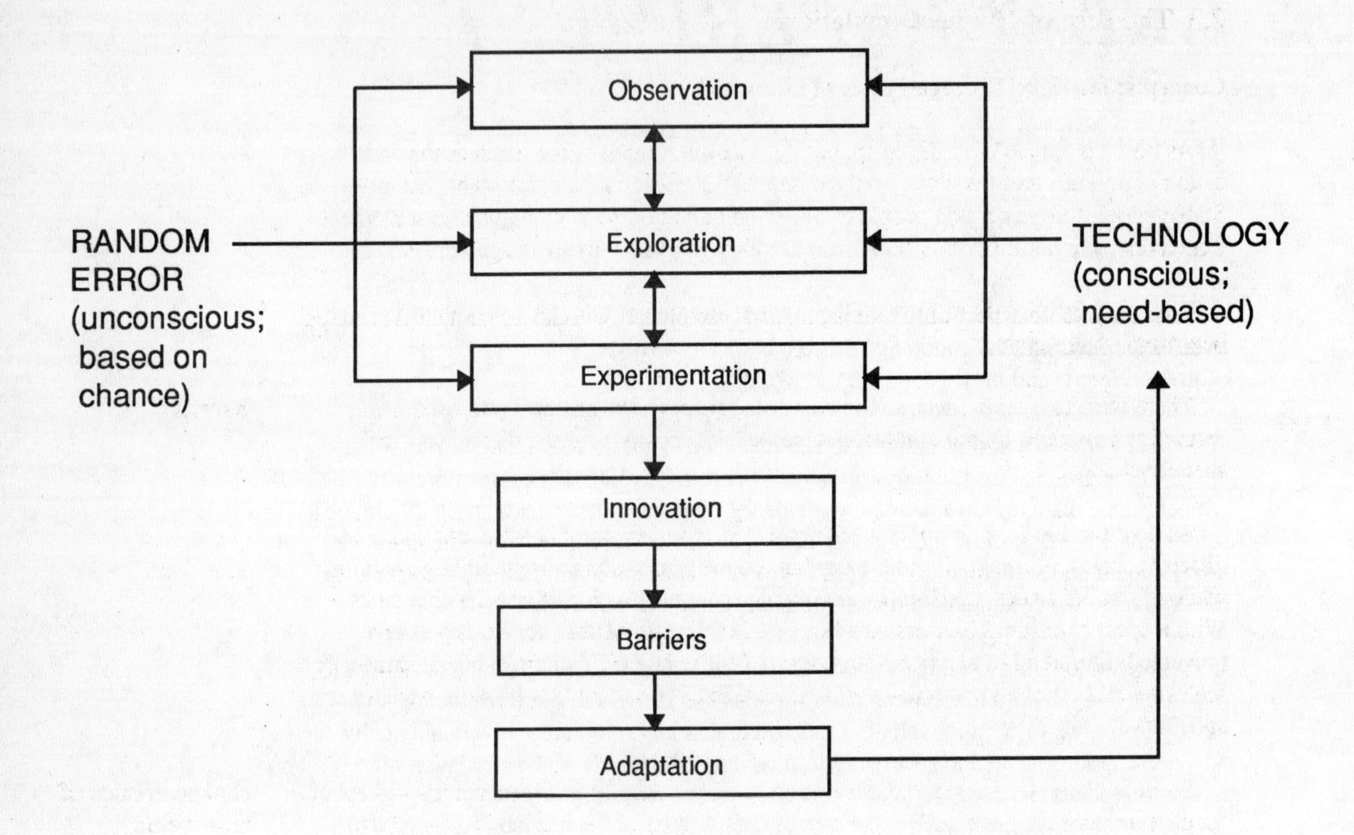

Figure 2.1 Process of human adaptation

In the diagram, the occurrence of the first three capabilities was not necessarily sequential. Experimentation may have preceded observation or exploration. Fire was likely observed first as a result of lightning striking the ground, or from volcanic eruptions. But humans could have experimented with stones or sticks first, then observed the effects of friction later.

innovation

Observation, exploration, and experimentation produce innovations—objects or ways of doing something which did not exist before. If the innovation was a suitable match for the need, its use became part of the learned behavior or culture of the people—it was adapted.

Random error also influences the capabilities of humans. It is possible that the benefits of fire were realized when lightning ignited an area close to people. The people were cold, felt the heat from the fire, and this caused them to seek a way to attain the same effect.

This whole process is constantly being recycled. After humans adopt and adapt to an innovation, the innovation itself may improve their capability to observe, explore and experiment. Telescopes and microscopes are good examples of technology, or applied innovations.

adaptation

2.3 The Rise of Human Populations

Concepts: barriers, characteristics of innovation

As humans spread over the earth, basic agricultural innovations, such as the domestication of animals and the plow, enabled human population to become stable and grow. Stable agriculture permitted people to make the transition from nomadism to settlement. Centers of agricultural production formed, followed closely by towns and cities.

What factors initially influenced the amount and rate of humans who spread into these new areas of settlement?

There were two main factors. The first was barriers—the physical and cultural restrictions imposed by the climate, landscape, and society; and the second was the specific characteristics of the innovations which were occurring.

barriers

As noted in *Environmental Geography*, physical and cultural barriers can operate at different scales. Extreme climates have limited the populations of several large regions, while at smaller scales steep slopes can prevent the inhabitation of areas within cities. With respect to cultural barriers, at a large scale religious beliefs, for example, have prevented the spread of alcoholic beverages to entire Islamic nations; while at smaller scales several cities and counties within the United States' "bible belt" were once, or are still, "dry."

Specific characteristics of the innovation itself can help spur its acceptance—in effect "sell" it to more people quickly, and over longer distances. Researchers have classified several factors which help push innovations. These are simplicity, observable benefits, relative advantage, ability to make small trials, and compatibility. We can examine the farm plow within this framework.

characteristics of innovation

Simplicity refers to the ability to explain and understand an innovation quickly. Early plows originated in Mesopotamia and Egypt, and were developed from digging sticks. Many of the first plows were little more than digging sticks just tied to an animal and dragged through the ground (Fig. 2.2).

Observable benefits allow potential users to see what they might be getting. It was easy to demonstrate that animals could pull plows over longer distances with much less human effort.

Relative advantage can refer either to the innovation having benefits greater than the costs or risk, or being easier to use with similar expense or risk. Plows were an easier and more productive use of the same tool; the risk was minimal since animals were domesticated.

Ability to make small trials allows users to demonstrate simple innovations quickly,

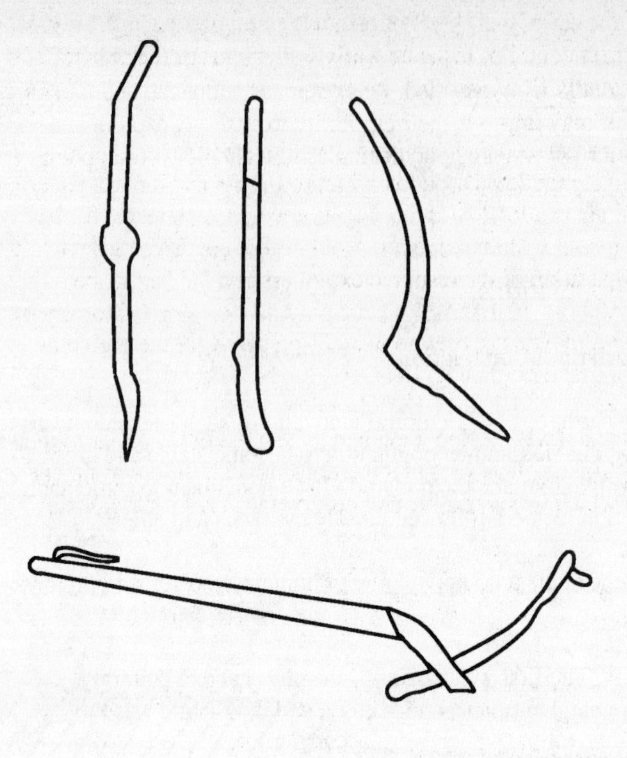

Figure 2.2 Early plows

and to test more complex innovations before expanding their extent. A plow could be demonstrated easily in your own back yard.

Compatibility refers to the fit of the innovation with current cultural practices. With food being a basic necessity, there were no religious or social restrictions to adopting the plow as a new tool.

We can think of the plow as a form of technology, an innovation which served to fulfill human needs. The wheel and stone tools are other early examples. As population grew and spread, technology was applied to wider areas; Sometimes the new technology itself allowed people to overcome physical barriers and travel longer distances in less time; e.g., the wheel.

One effect of these technological improvements was to increase human impacts on the landscape. For centuries agriculture disrupted original landscapes as forests were cleared and wetlands drained. Today, population expansion and economic development are threatening previously remote and lightly populated regions.

Is all technology harmful to the environment?

2.4 Frontier Environments and the Modern Threat

Concepts: frontier, mismatch of technology and the natural environment, native peoples

When early peoples settled the earth, their level of technology required a high level of dependence on the natural environment. For instance, early settlements of the Nile reflect a complete dependence on the yearly flood cycle of this river.

mismatch of technology and the natural environment

Today, when technologically advanced nations try to develop frontier areas, there is often a great mismatch between the methods used to extract resources and their impact on the environment. Huge disruptions result when technologies developed somewhere else are superimposed on an environment without adaptation and mitigation. This mismatch arises partly from the spatial separation of the resource exploiters and the landscape containing the resources. For example, British colonial operations were applied locally by people steeped in British values and landscapes who were ill-prepared for the problems of plantation farming or mining in the tropical regions.

Returning to the Nile, inhabitants who settled there had to stay there. Desert surrounds the river; the river is literally a lifeline. This society's relationship with the environment was based primarily on a need which was available and observable at a proximal scale (Fig. 2.3).

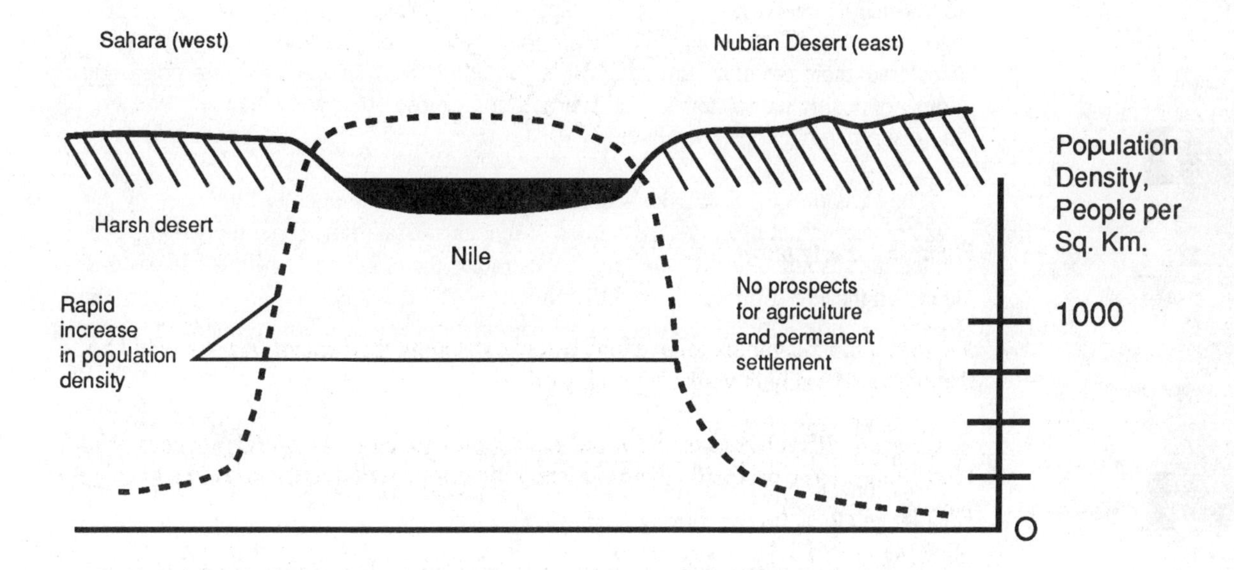

Figure 2.3 Nile River settlement pattern

On the other hand, oil and mining companies and their governments are at a distal scale to the natural environment they are using technology to exploit. And, the primary motivation is profit—which is their primary need—but not the sole need of the entire society. The result is a big mismatch between the land use practices, the environment, and the native peoples. Native peoples' cultures and livelihoods "in the way of progress" often bear the brunt of these unbalanced practices. For instance, the elimination of bison from the Great Plains helped to displace Native Americans from their land in the U.S. in the 1800's.

indigenous peoples

Is there a way to achieve sensible economic development accompanied by respect for native peoples? Vignette 2.1 addresses these issues.

Vignette 2.1 "Indigenous Peoples," by Ted Macdonald

"Whenever a shaman dies, it's as if a library disappeared." Such statements have been repeated and expanded through films like "Medicine Man" and countless articles. The images illustrate how the Western world allocates space in the environmental landscape—indigenous people provide us with botanical knowledge and holistic world views. As most astute and sympathetic observers agree, indigenous people manage their resources in a sophisticated and sustainable manner.

Yet many would lump indigenous people in with the flora and fauna—another "endangered species" to be preserved and observed in the name of biological diversity. While for some small, isolated societies, protection is essential for survival, to extend that response to all forest people overlooks a critical distinction between science and history—people evolve socially and politically as well as biologically. Thus, about 20 years ago, many indigenous communities, particularly those of Latin America, began to accelerate their evolution through political mobilization. These new, grassroots organizations are positioned to stem forest destruction far more effectively than any Western pleas to preserve indigenous knowledge.

The human rights link to environmental issues was dramatically illustrated by the 1988 assassination of Francisco "Chico" Mendes, a Brazilian rubber tapper who promoted the sustained use of standing forests rather than clearcutting for cattle pasture lands. In the eyes of many in the US and Europe, he was a conservationist. But to the local elites of the Brazilian Amazon, he was a labor organizer who threatened the status quo.

On a far grander scale, indigenous communities throughout the Americas and many other areas of the world have now organized themselves into local, regional, national, and international ethnic federations. Indigenous organizations in Bolivia and Ecuador recently challenged local powers through extended protest marches, and many of their leaders, like Mendes, risk violent consequences.

Where are the environmentalists in all this? Indigenous people form part of the forest landscape, and their culture—indeed their fate—is linked closely to that forest's future. Yet while some convergence of interests exists between long-term forest residents and those concerned with environment and development, the links are indirect and alliances are fragile. Most indigenous and other forest people recognize that any outsider can put their land and its resources—their present and future capital—at risk. So indigenous people are as wary of the term "conservation" as they are of "development," always asking: "For whom?"

In other words, indigenous people's primary concern lies with gaining secure land and resource rights, not just the complex management of that environment. They are often as suspicious of environmentalist as they are of development agencies—both groups often design plans for the use of the land and resources claimed by forest people. Park planners and managers in many countries of the developed as well as the less-developed world still cast indigenous people's rights alongside those of forest exploiters who must be restrained or expelled from protected areas.

Indigenous ethnic federations initially focused solely on political mobilization, an essential first step toward broader empowerment. Now many indigenous communities ask their organizations: "OK, we're organized. Now what?" These people are asking their leaders to pursue land claims and obtain technical assistance for programs of resource management.

Such requests enable those concerned with the environment and sustainable development to work directly with indigenous people as they draw from traditional technologies and create new ones. Support, however, must include respect for rights to land and resources. As illustrated in indigenous resource managed projects such as Panama's Project PEMASKY and Ecuador's Awa Project, indigenous people and other forest residents do not isolate technology, politics and culture.

Through their organizations, indigenous people have raised their heads and are demanding attention, not simply as objects of research, largesse or sympathy, but as legitimate actors on a broad political landscape. Given that attention, indigenous people will play a vital role in maintaining the world's resources.

The reading notes that indigenous peoples have become integral components of ecosystems, not just appendages of ecosystems. Indigenous peoples have developed legitimate roles as natural players in ecosystems such that their loss from the system would be as disruptive as the loss of any other "natural" parts. The reading also suggests that local interests at the proximate scale to the resources (indigenous peoples), have to be in the forefront of any attempt to achieve sustainable management practices. Only then will the sustainable use of the land by remote (distal) interests be possible.

What can be done to change the attitudes of environmentalists and developers toward indigenous peoples?

Finding a cure for cancer may depend upon the preservation of an unknown plant species in the wet tropics. This region and other frontier areas are currently under development pressure from internal and external interests. The next five sections discuss some of the problems and prospects for these regions.

2.5 The Wet Tropics

Concepts: limiting factors, leaching, erosion, ecological base case

limiting factors

All frontier environments have factors limiting their settlement. We usually think of water and temperature as being the main limiting factors. But the wet tropics have plenty of moisture, are not unbearably hot, and as a result have ample vegetation. What other factors have limited their settlement?

leaching

Poor soil is one factor. Heavy rains leach nutrients away from tropical soils. Thick vegetation and extensive root systems make the soil friable, or very permeable to water.

As water percolates downward, it removes nutrients from the soil, which are then transported away by groundwater and streams.

erosion

Excess water that cannot be infiltrated into the ground runs off and carries small particles of soil with it. This process, called erosion, often removes the organically rich top layer of soil; resulting in soils not well-suited to support agriculture.

Lastly, the favorable climatic conditions which have made the tropics so well suited for species diversity have made it tougher for human habitation. Many species of insects, snakes, and parasites harmful to humans thrive in the tropics. Each year, 200-400 million new cases of malaria occur, causing 1-2 million deaths.

ecological base case

These limitations on settlement have, despite modern development pressure, preserved the natural environment in some tropical areas. This is allowing scientists to study the interrelationships of nature in their true form—an ecological base case providing models for future management of rainforests and other tropical environments. Environments affected by human disturbance often hinder the study of the original relationships. For instance, as we try to understand the migratory patterns of whooping cranes, it is complicated by the disruption of their water supplies resulting from grazing and agriculture.

2.6 The Dry Lands

Concepts: desertification, irrigation efficiency

desertification

Desertification begins with land that has marginal capability to sustain agriculture. The land has poor soil, highly variable moisture inputs from year to year, and often experiences high erosion rates from wind.

People come next, and through inefficient farming methods, do not allow the land to recover its scarce resource base of soil moisture and grass cover after a period of use. Overuse of the land often coincides with periods of drought, and this causes a severe reduction in the land's productive capability. Each year on the earth's surface an area equivalent in size to the state of Nebraska undergoes desertification.

Much of the damage could be avoided through better land and water management techniques. Erosion control practices are a good place to start, since certain measures can improve land and water conditions simultaneously. For instance, planted vegetation can help anchor the soil, and it also slows the rate of water flowing over the surface, thus reducing erosion of the most fertile layer of soil into nearby waterbodies.

irrigation efficiency

Irrigation efficiency is another area where improvements can be made. Worldwide, the efficiency of irrigation systems is estimated to be only 37 percent. That is, only a little more than one-third of the water applied to irrigated fields actually reaches the plants--the rest is lost to evaporation and groundwater.

Figure 2.4 shows some of the irrigation techniques used. Flood and furrow irrigation has efficiencies between 40-60%; sprinkling systems are 60-75% efficient; and trickle irrigation efficiencies approach 90%. The wide ranges of efficiency with each technique underscores the importance of proper management. For instance, savings of up to 30% from flood and furrow irrigation can be realized if the excess runoff is recycled back into the field.

Flood and Furrow

Center Pivot Sprinkling System

Trickle System

Figure 2.4 Irrigation techniques

In arid lands, drip irrigation has been the most successful since the initial water requirements are the lowest. Seepage and evaporation are minimized. However, this technology costs about $1000 per acre, which is much too expensive for many of the less developed nations to afford.

A low-tech and low cost variation of drip irrigation uses baked earthen pitchers filled with water. Vegetable or fruit seeds are planted around the pitchers, which have slightly porous walls that allow water to seep slowly out. This technique shows promise for certain crops, especially pumpkins and melons.

Can current technology alone stop desertification? To answer this, we should first consider the limitations of certain landscapes, and accommodate the frequency of our land use to these constraints. We also need to change our economic systems to ones in which variability in productivity is the norm, i.e., abandon the simplistic steady flow assumption.

2.7 The Cold Lands

environmental hazards

Concepts: environmental hazards, magnitude and frequency

Cold creates special environmental hazards. Water expands when frozen, and this can cause frost heaving and uneven sagging of the ground (Fig. 2.5)

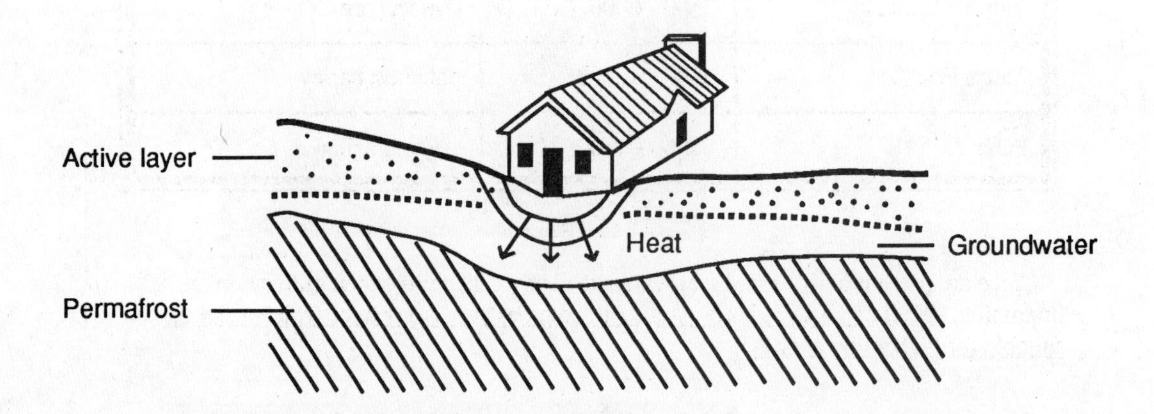

magnitude and frequency

Figure 2.5 Hazards in cold climates

Human activity can increase the magnitude and frequency of environmental hazards. Magnitude refers to the energy released by the hazard and the extent of the damage to humans—loss of life and property. Frequency is the rate of occurrence—the number of times a hazard occurs in a fixed period of time.

As Figure 2.5 shows, buildings give off heat, which can change the annual rhythm of heat flow in and out of the soil. If building density increases, then more heat is applied to the same area, resulting in more frequent and damaging soil disturbances.

Much like the desert areas, cold lands exhibit the abrupt transitions between developed and undeveloped areas; e.g., the population distributions of Saskatchewan and Manitoba.

2.8 The Mountain Lands

Concept: altitudinal zonation

As we gain altitude, air temperature declines at a rate of 3.5° F. per thousand feet. This decrease in temperature (called the lapse rate), and the rugged terrain are the main limiting factors in mountainous regions. As a result, growing seasons are shorter, and as the ground gets steeper and elevation increases the soils become thinner.

altitudinal zonation

Distinct land use zones are formed due to the constraints of the climate at different elevations. This process is called altitudinal zonation. Table 2-1 provides an example of this occurring in Central America.

TABLE 2-1 Altitudinal Zonation in Central America

Zone Name	Elevation (feet above sea level)	Primary Land Use
Tierra Caliente	0 - 2500	plantations (bananas)
Tierra Templada	2500 - 6000	coffee, corn
Tierra Fria	6000 - 12000	potatoes, barley
Puna	12000 - 15000	sheep grazing

Crops are present until 12,000 feet, where the snowpack makes their cultivation impossible. What characteristics do you think potatoes and barley have which make them suitable for higher elevations?

2.9 The Continental Shelves

Concepts: common resources, environmental equity

The continental shelves are the submerged continuations of the coastal plains of continents. Although not habitable, they possess great species diversity and mineral wealth. Continental shelves have been the focus of intense economic activity including commercial fishing and mineral exploration.

The reasons for this interest is partially due to the adjacency of the area to major concentrations of population. Despite the diffusion into frontier lands, over half of the world's population still lives within 100 kilometers (62 miles) of the sea coast.

Two critical issues concerning this region are: Who owns the rights to the natural wealth there, and how large are the territorial rights of each country?

common resources

Nobody "owns" the oceans—they are a common resource. As the value of the continental shelves became known, nations wanted a means to harness that wealth near their shores. Several nations began to claim territorial rights—areas under their jurisdiction extending outward from their coastlines a given distance into the sea. This claim allowed them to restrict passage of foreign vessels, thus eliminating any chance "their" fish or oil would be caught or extracted by another nation.

Things heated up as nation after nation expanded their claims to adjacent waters. In 1995, Canada and Spain had a dispute over fishing rights in the North Atlantic. Table 2-2 traces the major events in the history of territorial claims.

TABLE 2-2 Division of the Seas

Year	Status
1945	40 nations had a 3 mile territorial limit (TL)
1982	United Nations Conference on Law of the Seas (UNCLOS). This created exclusive economic zones (EEZ), which provide for control of mining and fishing to 200 miles offshore.
1990	12 nations had a 3 mile TL; 99 had a 12 mile TL; 11 had a 200 mile TL
199?	World Lake concept - Control for each nation is extended to median lines in the seas (Japan and the USA would be neighbors under this arrangement).

Controversy over territorial rights continues. How do the less-developed nations--which do not have the technology to take advantage of coastal or deep sea resources—reap some of the benefits of this common resource? And, what arrangements are made to provide access for landlocked nations?

environmental equity

The questions above address the issue of environmental equity, or fairness. Is there a way to create a fair distribution of the resources for poor and landlocked nations?

Applications / Discussion

2.2) You need at least 10 people for this application.

To illustrate diffusion and barriers, whisper the phrase "we don't spend enough money on environmental research" to the first person in the group. Then instruct that person to whisper what he/she heard to the next person. Continue until the last person has heard the phrase. When completed, ask the first and last receivers of the message to write what they heard on the blackboard.

Discuss the reasons for any differences in the two phrases. Were barriers present? Did some people adapt the message to their values?

2.3) How does recycled paper fit the criteria of simplicity, observable benefits, relative advantage, ability to make small trials, and compatibility? What are the potential barriers to using recycled paper?

2.4) Using the data for population density in Appendix A, construct a population density profile similar to Figure 2.3. At the margin of the urban and frontier environments you have constructed, what could cause the sharp transitions to shift, and why?

2.5) Observe the differences between the soil at the edge of a heavily wooded area and 100 yards away on the adjacent land. How might water flow and erosion differ between both areas?

2.6) When you water the lawn, what factors affect the irrigation efficiency?

2.7) Should we work harder to develop technical solutions which allow us to build structures in cold climates, (assume some risk of frost heaving remains), or try to persuade developers and settlers to stay outside of these marginal lands? What are the benefits and drawbacks of each approach?

2.8) Identify some of the special problems for human settlement related to water supply at higher elevations.

2.9) In 1995 Canada and Spain had a dispute over fishing rights in the North Atlantic. A Spanish ship was fishing outside of Canada's 200 mile territorial claim, yet warning shots were fired across its bow by a Canadian patrol boat. Canada claimed the reserves of fish were so depleted that any fishing in the area would threaten the resource. Spain argued that it had a right to fish in an area not "claimed" by any nation—its fishing industry has recently been hurt by other nations' claims to large ocean areas. Should broader environmental concerns (e.g., the preservation of the world's fish stocks) take precedence over individual nations' legal claims and economic need?

Recommended Reading

Burke, J. (1985). *The Day the Universe Changed.* Little Brown, Boston, MA.

Meggers, Betty J. (1971). *Amazonia: Man and Culture in a Counterfeit Paradise*. Aldine, Atherton, Chicago, IL.

Sources of Additional Information

Chihuahuan Desert Research Institute, P.O. Box 1334, Alpine, TX 79831

INTERNET: Drought Action News. University of Nebraska, 236 Chase Hall, P.O. Box 830728, Lincoln, NE 68583-0728
e-mail: agme@unlvm.unl.edu

United Nations Environmental Programme, Room DC2-803, United Nations Plaza, New York, NY 10017

World Resources Institute, 1709 New York Ave, NW, Seventh Fl. Washington, D.C. 20006

References

Curwen, E.C., and G. Hatt. (1953). *Plough and Pasture: the Early History of Farming*. Collier Books, New York, NY.

Deshmukh, Ian. (1986). *Ecology and Tropical Biology*. Blackwell Scientific, Palo Alto, CA.

Glautz, M.H., ed. (1977). *Desertification: Environmental Degradation in and Around Arid Lands*. Westview, Boulder, CO.

Postel, S. (1985). *Conserving Water: The Untapped Alternative*. Worldwatch Paper #67, Washington D.C.

Rogers, E. (1983). *Diffusion of Innovations*. 3rd ed. Free Press, New York, NY.

United Nations Office of Legal Affairs. (1993). *The Law of the Sea: definition of the continental shelf*. New York, NY.

Webb, W.P. (1986). *The Great Frontier*. Univ. of Nebraska Press, Lincoln, NE.

3

GLOBAL CYCLES AND SYSTEMS: THE ENVIRONMENTAL FRAME-WORK

Review

3.1 Summary

As the life support systems on earth developed—the atmosphere, lithosphere, and hydrosphere, new distributions of matter and energy were created. These new arrangements, coupled with favorable amounts of incoming solar radiation, allowed for the initiation and sustenance of living matter on earth and formation of the biosphere.

The biosphere, hydrosphere, lithosphere, and atmosphere are the major divisions within the geosphere, or total earth environment. Balances of matter and energy moving through the hydrosphere, lithosphere, and atmosphere are necessary to sustain the biosphere, which exists within a narrow zone intersected by these three spheres.

As energy moves through the earth environment it follows heat and organic pathways. The heat path of solar energy drives the movement of air, water, and nutrients at and near the earth's surface. The organic path of solar energy begins with photosynthesis, which creates the necessary organic matter to support food chains of organisms. Ecosystems result from the activities which compel plants and animals to survive.

Earth's environment thus functions by complex interactions of energy and material systems. Each system is made up of chains of cause and effect relationships, and a dynamic balance is achieved by positive and negative feedback within the context of global energy and matter cycles.

3.2 The General Organization of the Earth's Environments

Concepts: four spheres, interrelationships, matrices

four spheres

Natural landscapes develop in the geographical sphere encompassing the Earth's crust (lithosphere), the zone of water (hydrosphere), the lower atmosphere up to the ozone layer, and the zone of living matter (biosphere). These layers overlap one another and are all closely interrelated.

Flows of energy and matter constantly traverse each sphere or zone. As a result, any change to a process in one sphere—through its own development or from humans, leads to changes in the other spheres.

interrelationships

We can construct a table to show the scope of some common human activities within the context of the four spheres. This provides a useful framework for identifying interrelationships in the environment (Table 3-1).

TABLE 3-1 Human Activities and the Four Spheres

Sphere	Human Activity		
	1. Lawn Mowing	2. Fertilizing	3. Raking Leaves
1. Lithosphere	yes (1,1)	yes (1,2)	yes (1,3)
2. Biosphere	yes (2,1)	yes (2,2)	yes (2,3)
3. Hydrosphere	yes (3,1)	yes (3,2)	yes (3,3)
4. Atmosphere	yes (4,1)	yes (4,2)	yes (4,3)

matrices

The arrangement of data in row and column format as shown here is called a matrix. The first row element is "lithosphere" and the first column element is "lawn mowing." Where each row and column element intersect is called a cell. The cells have been numbered by their row and column position.

In cell (1,1) where lithosphere and lawn mowing intersect, is there a relationship (a process or impact) between the lithosphere and lawn mowing as a result of this activity? At the most basic level we answer "yes" or "no".

It seems that every activity related to normal lawn maintenance is connected in some way to every sphere. Let's identify a connection for each cell:

Lawn Mowing (cells 1,1 - 4,1)

(1,1) grass clippings decompose and help regenerate the organic layer of the soil

(2,1) soil organisms decompose the grass clippings

(3.1) more water is kept on the lot after it rains; grass clippings slow down runoff and allow more water to infiltrate into the soil; if clippings are removed, shorter grass is less effective in slowing down runoff and inducing infiltration

(4,1) lawn mower engines generate large amounts of air pollution, largely because they lack pollution control devices

Fertilizing (cells 1.2 - 4.2)

(1.2) most fertilizers add the nutrients nitrogen, phosphorus, and potassium; this affects the chemical balance of the soil

(2.2) some fertilizers are harmful to birds

(3.2) after rains, fertilizer is carried with runoff into storm drains; eventually making its way into a lake or stream. In freshwater lakes, excessive amounts of nutrients—such as the phosphorus in fertilizers, can stimulate the production of algae—this is part of a long-term process called eutrophication.

(4.2) sprayed fertilizer often gets airborne and can pollute the atmosphere

Raking Leaves (cells 1,3 - 4,3)

(1,3) leaves provide a good source of organic nutrients for the soil; raking them prevents this

(2,3) raked leaves often end up in the storm drain. If the leaves make it to a stream, insects such as stoneflies and caddisflies shred the leaves and speed up their decomposition

(3,3) if the leaves make it to a stream used for a community's water supply, excessive amounts of organic matter can discolor the water and raise treatment costs

(4,3) burning leaves pollutes the atmosphere

We could refine this technique by discriminating between normal processes, beneficial impacts, and harmful impacts. Instead of using yes/no in the cells, we might use a "0" to indicate the normal processes and beneficial impacts, and "2" for harmful impacts. Coding with numbers allows us to obtain the sums of the rows and columns.

Suppose we use the coding scheme above and code the processes and impacts of lawn maintenance activities on the four spheres. What would higher row and column totals indicate (Table 3-2)?

3.3 The Source of Energy for the Earth's Environments

Concepts: energy balance, radiation/moisture relationships

energy balance

At the earth's surface, the sun is virtually the sole source of energy. As solar radiation heats the earth's surface, the amount of energy emitted is a function of the earth's temperature. If an area receives more energy, its surface temperature rises and more energy is in turn emitted. If less energy is received, surface temperatures fall and less energy is subsequently emitted. Therefore, if energy is received at a constant rate over a long period of time, constant rates and temperature and emittance develop. This balance of energy inputs and outputs is called the earth's energy balance, and the resultant temperature is the equilibrium surface temperature (Fig. 3.1). For the earth as a whole,

TABLE 3-2 Lawn Maintenance: normal processes and harmful impacts (0=normal process, 2=harmful impact)				
Sphere	Human Activity			
	Lawn Mowing	Fertilizing	Raking Leaves	Total
Lithosphere	0	2	2	4
Biosphere	0	2	2	4
Hydrosphere	0*	2	2	4
Atmosphere	2	2	2	6
Total	2	8	8	

*** assuming the grass clippings are left on the lawn**

the current equilibrium surface temperature is 15° C. (about 50° F.)

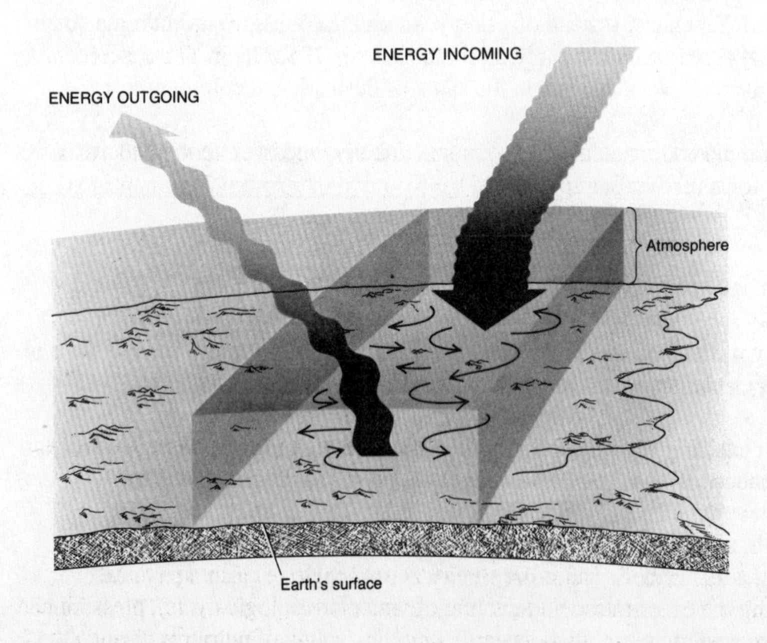

Figure 3.1 Earth's equilibrium surface temperature

Existence of a long term energy balance at the earth's surface does not mean energy is distributed exactly the same at all locations on the earth. The curvature of the earth creates general zonal distributions of energy. The high latitudes (66.5-90 degrees latitude in both hemispheres) receive much less energy than the low latitudes (0-23.5 degrees latitude), and the midlatitudes (23.5-66.5 degrees latitude) receive an intermediate amount. Due to cloud cover, it is in the subtropics (20-30 degrees latitude) where the greatest amount of solar energy is received.

radiation/moisture relationships

The distribution of moisture is also broadly zonal over the earth. Since temperature and precipitation are the key variables of climate, a general classification of climate can be developed from the coincidental distributions of solar radiation and moisture (Table 3-3).

TABLE 3-3 General Climate Classification

Solar Energy Input	Moisture Availability	
	High	Low
High	tropical	subtropical
Low	temperate	polar

At the most basic level, there are two zones of energy and moisture availability (high and low), which yield four possible climatic combinations. The cells in Table 3-3 identify the resulting climates.

This is a general classification at a large scale. Globally, and over very small areas, energy flows and forms are constantly changing.

3.4 The Major Energy Systems of Earth

Concepts: energy paths, conservation/conversion of energy (first and second laws of thermodynamics), energy states (potential/kinetic)

Solar radiation entering the earth's environmental systems follows one of two paths—heat or organic (chemical). Over 99% of solar energy follows the heat path.

energy paths

In the heat path, solar energy is taken up by the atmosphere, oceans, or continents. Once resident in the air, land, or water the energy is used to drive earth's physical processes including the evaporation and cycling of water (hydrologic cycle), pressure and wind systems (atmospheric circulation system), and the cycling of nutrients through the water, soil, and air (biogeochemical cycles).

The organic path of energy begins with photosynthesis, the process by which plants convert solar radiation into plant materials. Food chains of organisms then pass on the organic materials manufactured by plants. Each successive food chain depends upon one or more previous food chains for its existence. Life thus rests on a foundation of solar energy.

conservation/ conversion of energy (first and second laws of thermodynamics)

As energy is used it changes form. "Use" might be a misleading term, because energy is never used up in the sense that some of it disappears. Energy is the capacity to do work; and the capacity to do work is always conserved regardless of the energy form; energy is never created or destroyed. This is the first law of thermodynamics. When energy changes form, some of the efficiency of the energy is "lost." This is the second law of thermodynamics.

<u>Is the energy lost really "lost</u>?" No. The energy "lost" by conversion is conserved as heat, which is another form of energy. This is the first law of thermodynamics in action again.

energy states (potential/kinetic)

There are two major states of energy: potential and kinetic. Matter at rest has potential energy; matter in motion has kinetic energy. Conversion between these forms of energy often occurs continuously, as in a dam.

The water held stationary in a reservoir behind a dam has the potential capacity to do a certain amount of work, such as driving a turbine. When the water is subjected to a gravitational gradient, (such as the sharp drop over a dam), the potential energy of the stored water is converted into the kinetic energy of the falling water. Some energy is lost as friction as the water moves over the spillway and through outlet pipes, but this friction generates heat, so all energy is conserved.

3.5 The Cycle of Matter in Ecosystems

Concepts: ecosystems, biogeochemical cycles, sources, sinks, fluxes

ecosystems

Ecosystems are energy systems made up of different groups or levels of organisms. Each level of organisms depends on the previous level (or levels) for its energy (food) supply. Organisms convert food into energy; a small part of this is conserved as body mass, and the rest is given up to the atmosphere through respiration; heat released in breathing, for example. Applying the second law of thermodynamics, we would expect that as a result of these conversions, the supply of available energy in ecosystems decreases very rapidly as you ascend the food chain. It does.

biogeochemical cycles

Other forms of matter, or nutrients, are required to support ecosystems. As the atoms and molecules of nutrients move through ecosystems matter is conserved. Thus, the biogeochemical cycles are great recycling systems for nutrients like carbon, oxygen, nitrogen, and phosphorus.

sources, sinks, fluxes

It helps to conceptualize biogeochemical cycles as having three components: sources, sinks, and fluxes. Sources are emitters, and sinks the absorbers of the nutrient. The rates of flow between the sources and sinks are the fluxes. Sinks can also be sources of the nutrient; the atmosphere is sink and source for carbon in the form of carbon dioxide (Fig. 3.2).

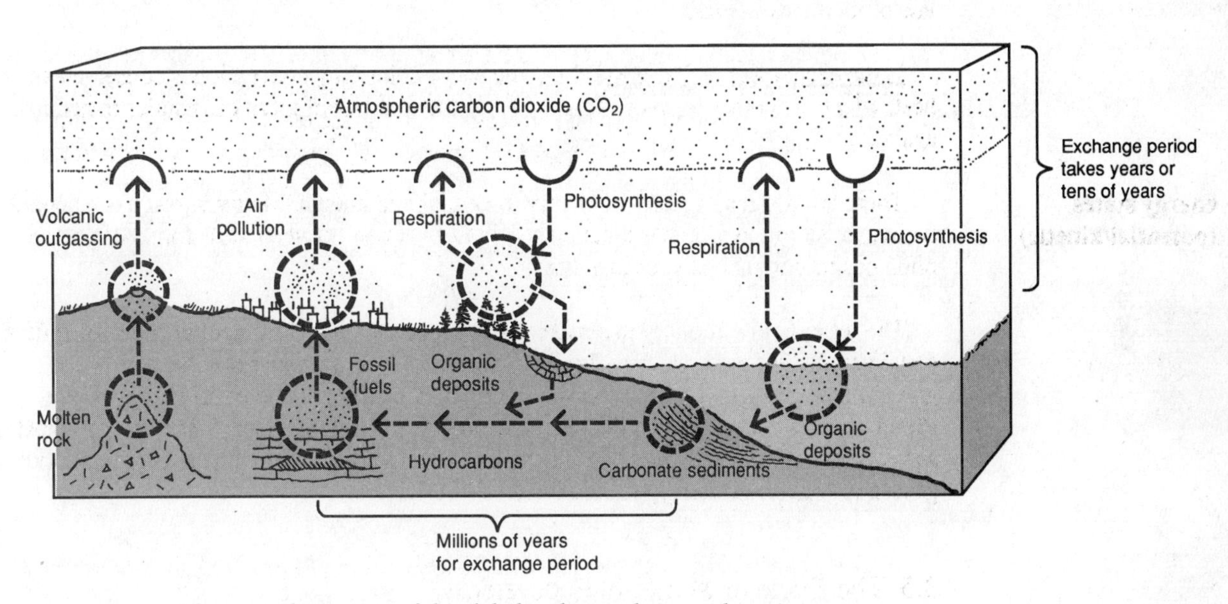

The general structure of the global carbon cycle in modern times.

Figure 3.2 The carbon cycle

3.6 Interrelations in Environmental Systems

Concepts: causation, feedback

Complex interactions between the organisms maintain the balance of matter and energy in ecosystems. These interactions are of two types: primary (direct cause and effect), and secondary (feedback).

causation

Due to the complexity of environmental systems, causation (cause and effect) is often difficult to determine. We are always asking: What caused that?

There are generally three criteria used to define "what caused that": 1) the cause precedes its effect in time; 2) cause and effect and empirically related, i.e., observed and/ or measured; and, 3) the relationship is not due to the presence of a third variable. All three criteria should be present together, and if so, you can begin to feel more confident that "what caused that."

The requirement that cause precedes effect is a lot easier to determine when you can observe the events. Since none of us were around when the solar system was created, it took a long time to figure out how the planets were formed. Conversely, if we throw a stone into a pond and waves are generated, we see the cause and effect clearly.

An empirical relationship is one that is based on observation, and consists of events that can be verified or disproved by experimentation. Duplicating an experiment many times can help to determine cause and effect, but you cannot rely only on experimental replication. If the first experiment failed to eliminate a third, or other variables, then the error will be repeated in the replications. Sometimes, different results show up later. Harmful side effects in people after a drug has been lab tested and approved provide examples of how empirical evidence is often not sufficient to determine reliable cause and effect relations.

An example of a relatively simple process will help demonstrate the procedures required for determining cause and effect.

You are certain the absence of light can cause retarded growth in a certain plant species, but a friend of yours is skeptical. How do you show that light is the cause of the retarded growth?

You design an experiment. Two identical pots containing the same plant species (of very young age and size), and soil type are used. One plant receives no light, the other one does. In one week the plant which has not received light is 2 inches shorter than the other.

<u>Have you proven anything yet, that is, have you defined the cause of the diminished growth?</u>

Your friend says no, because you have only done one trial, and the growth retardation could have been caused by the presence of another variable, such as a hidden root disease or water supply. You have only demonstrated part of the first criteria: that growth lags might follow "something."

The experiment is repeated as before, and the amount of water each plant receives is identical. You also take great pains to eliminate other possible influences (the "third" variable), such as temperature variations and wind currents within the lab. After 20 more trials, the results are the same. You have started to demonstrate an empirical relationship between the amount of light and growth, and other variables do not seem to be a factor.

Two more events would help a scientist say with confidence that for this species, if light is lacking, growth will be retarded: 1) replication of the same experiment by other labs; and, 2) many observations of the same phenomenon in the field by different investigators.

Imagine how tough it is to prove cause and effect when you do not have control over all the variables. This is often the case with environmental disturbances caused by human activities. Human activities are normally complex. For example, land clearing leading to species changes may involve cutting, burning, soil scraping, chemical applications, and/ or the introduction of invasive weeds, insects and other organisms.

feedback

Sometimes, the primary (first) effects generate additional effects. These secondary effects are called feedback. Auto exhaust and industrial smokestacks have the primary effect of raising the amount of small dust particles in the air. The increase in particles may generate more frequent storms, since small particles such as these are required for droplet formation before precipitation occurs. This increase in storm frequency is one form of positive feedback; photochemical smog would be another.

3.7 The Gaia Concept

Concepts: Gaia hypothesis, hypotheses

The identification of the four spheres of the geosphere provides a general framework for studying major environmental interactions on earth. Currently, scientists are debating how the activities within each sphere affect life. The conventional view holds that life is dominated by the geophysical environment, and life is more or less subject to whatever turns the environment happens to take.

Gaia hypothesis

hypotheses

In contrast to this view, the Gaia concept states that life is able to maintain the environment for its own good. The Gaia hypothesis has three basic ideas: 1) life affects the physical environment; 2) life helps to stabilize the environment, such as the atmospheric heat balance and the cycling of matter; and, 3) the earth is an organism which can regulate itself (geophysiology). These ideas are hypotheses; a hypothesis is a tentative assumption made in order to test its logical or empirical conclusions. In order for a hypothesis to be valid, it must be testable, and represent something disputable. For instance, if somebody puts forward the hypothesis that people can fly, it is disputable and easily tested.

Scientists generally accept the first statement of the Gaia hypothesis, and some accept the second. The idea that earth, without the influence of outside forces, can and does regulate itself is not accepted because it cannot be tested. We have not seen, nor have we identified all of the possible interactions within ecosystems yet, and as a result we do not know all of the consequences.

Another problem with the Gaia hypothesis is the failure to account for objective geophysical processes, such as volcanic activity, as a co-shaper of the environment.

Applications / Discussion

3.2) From the list below, identify one activity which does not have a direct connection to either the atmosphere, lithosphere, biosphere, or hydrosphere.

crop farming; coal mining; logging; TV communications; thinking; war; volleyball; hunting

3.3) Using figures a and b below, determine what the radiation/moisture relationships are in your region. How would you classify the climate? Next, using figure b (precipitation) find another area at the same latitude that has at least a 10 inch difference in average annual precipitation. Assuming the solar radiation inputs are similar, what other factors might account for the differences in the yearly amounts of precipitation?

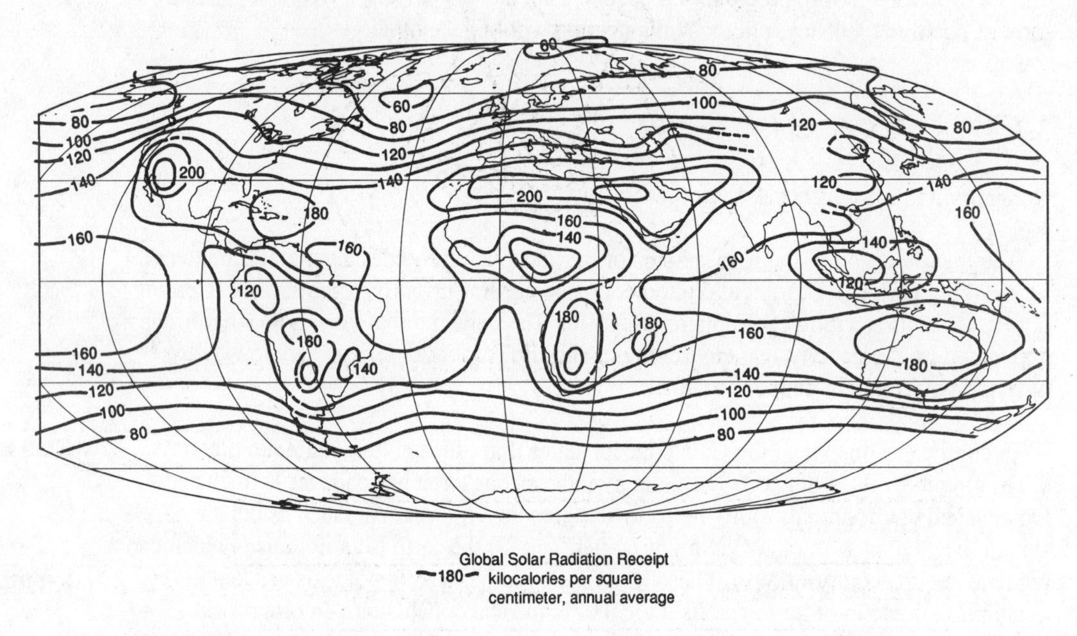

a) Global solar radiation inputs

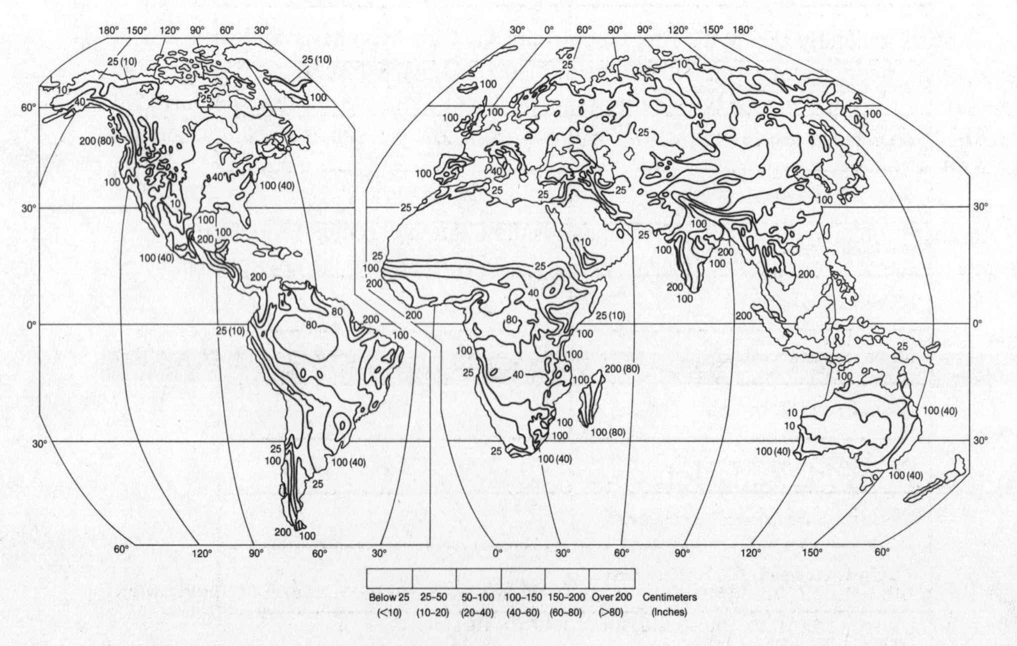

b) Global precipitation distributions

3.4) Discuss how energy from the chemical and heat paths interact when you ride a bicycle.

3.5) Given the table of carbon fluxes below, calculate the net flux for the atmosphere, land, and oceans.

Carbon Fluxes	
Direction of Movement	Flux (Gigatonne/yr)*
ocean to atmosphere	102
atmosphere to ocean	105
ocean surface to deep waters	39
deep waters to ocean surface	37
ocean surface to ocean life	28
ocean life to ocean surface	29
soil to atmosphere	60
life on land to soil	60
life on land to atmosphere	50
atmosphere to life on land	110
deforestation to atmosphere	1.6
fossil fuel combustion to atmosphere	5.4
* a gigatonne equals 1000 million tons	
Source: Rosenthal, (1995).	

For example, the net flux in gigatonnes for the atmosphere is:

ocean to atmosphere	+102
atmosphere to ocean	-105
soil to atmosphere	+60
life on land to atmosphere	+50
atmosphere to life on land	-110
deforestation to atmosphere	+1.6
fossil fuel combustion to atmosphere	+5.4
	+4.0

Compute the other net fluxes. What secondary feedback effects are possible from the extra carbon (as carbon dioxide gas) in the atmosphere?

3.6) A secondary effect of deforestation at the local level is the reduction in the amount of canopy, or vegetative cover, in the upper levels of the forest. As a result, more sunlight reaches the forest floor. What other local secondary effects can be generated by this activity?

3.7) How could you test the first part of the Gaia hypothesis: life (including humans) affects the physical environment? What data would you look at? What methods could you use?

Recommended Reading

Commoner, B. (1977). *The Poverty of Power*. Bantam Books, New York.

Ferris, T. (1988). *Coming of Age in the Milky Way*. Morrow, New York, NY.

Sources of Additional Information

INTERNET: Envirolink Gopher maintains an index of environmental organizations

Morning Sun Press. (1995). *Earthscore: Your Personal Environmental Audit and Guide*. Lafayette, CA.

Rosenthal, D.B. (1995). *Environmental Science Activities*. Wiley, New York, NY.

References

Bolin, B. and R.B. Cook, eds. (1983). *The Major Biogeochemical Cycles and Their Interactions.* Wiley, Chichester, England.

Lovelock, J.E. (1982). *Gaia: A New Look at Life on Earth.* Oxford University, New York.

Marsh, W.M. and J. Dozier. (1981). *Landscape: an Introduction to Physical Geography.* Wiley, New York, NY.

Odum, E.P. (1971). *Fundamentals of Ecology.* 3rd ed. W.B. Saunders, Philadelphia, PA.

Schneider, S.H. (1990). "Debating Gaia," *Environment* (32), 4-9, 29-32.

4

ECOSYSTEMS AND THE BIOCLIMATIC ENVIRONMENT

Review

4.1 Summary

Ecosystems function as energy conversion and transport systems for all life on earth. Beginning with photosynthesis, solar radiation is converted into chemical energy and then transferred through the ecosystem via food chains. The levels of transfer, or trophic levels, consist of organisms defined by their function in the ecosystem: producers, primary consumers, secondary consumers and tertiary consumers.

Climate largely determines the global distribution of ecosystems. Biomass production (productivity) and species diversity (biodiversity) are important barometers of ecosystem health.

Ecosystems have complex internal dynamics which reflect the interactions of organism communities. Change is continuous, and may occur more or less gradually through succession or suddenly by disturbance. Human interferences have altered ecosystems' spatial (geographic patterns) and internal operations (temporal dynamics), with the most noticeable changes on terrestrial ecosystems.

4.2 Basic Form and Function of an Ecosystem

Concepts: interrelationships, food chains

interrelationships

Everything within an ecosystem is interrelated. Ecosystems, even relatively "simple" ecosystems exhibit a large variety of interrelationships. If you disturb the integrity of the food web, then a complex variety of side effects can result.

food chains

Organisms higher on the food chain, e.g., bears, have several options available for their diet. When one source of food is removed, they can look elsewhere. But there is an impact—not only on the ecosystem originally impacted through the removal of food—there are also impacts on adjacent ecosystems (more bears in a new place).

Now consider human activity. Because of our requirements for food, clothing, and shelter, we draw off significantly more resources from ecosystems than animals. When we lose a specific source of food or fiber, we can often find it elsewhere, but at a greater cost. An example of this is our encroachment into the frontier lands, where the extraction of resources comes at a higher marginal cost. It requires longer distances and more time to transport the good, as with the Alaskan pipeline. Time is money.

This is one reason why it is so important to sustain habitats and biodiversity. In the long run, sustainable environmental practices are beneficial to economic development since they require efficient methods of resource use.

4.3 Plant Productivity: Feeding Ecosystems

Concepts: productivity, limiting factors, optimum conditions, tolerance thresholds

productivity

Productivity in terrestrial or aquatic ecosystems is measured by the amount of organic matter added to the landscape, lake bed, or ocean floor. The amount of organic matter produced depends on the available resources for photosynthesis and the magnitudes and frequencies of external disturbances.

limiting factors

Five basic resources are required for photosynthesis: light, heat, water, carbon dioxide and nutrients. Any of these basic resources can become a limiting factor in photosynthesis, either through shortage or oversupply. External disturbances, such as water pollution, can create conditions of shortage and excess simultaneously. This sometimes happens, despite good intentions when we try to protect wetlands.

optimum conditions

tolerance thresholds

Wetland ecosystems have species of plants whose optimum conditions for existence occur within certain water levels. In some new housing developments where the goal is to "save" wetlands, the storm drains from streets and yards lead directly into a wetland. In heavy rains, the water levels of the wetland rise significantly, and the tolerance threshold of the plants is exceeded. If the new volumes of water contain phosphorus, thick mats of algae may form, limiting the light available to organisms below the surface.

4.4 Types of Ecosystems and Their Characteristics

Concepts: biomes, biogeography, species diversity (biodiversity)

biomes

There are five major types of terrestrial ecosystems: forest, savanna, grassland, desert, and tundra; these are classified on the basis of the structure and composition of their vegetation. Within each major class there are distinct regional combinations of vegetation and animal life associated with a general climate type. These biogeographic areas are called biomes. Table 4-1 shows the general characteristics of earth's principal biome—the forest biome.

biogeography

The information in Table 4-1 demonstrates a basic principle of biogeography: harsher climates have fewer species with larger populations. The boreal forests contain the smallest trees and are the lowest in biomass production, but are the largest biome in terms of geographic extent, stretching east-west across the subarctic of North America and Eurasia. Continuing down the table, as the availability of heat and moisture increases, biodiversity increases in the forest biomes, but the extent of their core areas decreases.

species diversity (biodiversity)

A simple measure of biodiversity (computationally simple, but not necessarily easy to prepare) is to consult historical records and list the total number of species which have inhabited a given area since humans were present (you need to consult the fossil records). The number of species currently present in the area is then recorded. The ratio of the

TABLE 4-1 Biogeography of Forest Biomes

Forest Biome	Climate	Productivity	Extent of Core Area	Biodiversity*
Boreal	cold, low moisture	low	wide	low
Temperate	mild, moderate moisture	medium	medium	medium
Tropical	warm, very moist	very high	narrow	very high

* biodiversity is the total number of species in a given area

total number of species currently present to total number of species which have been present is the biodiversity index:

Total # of species currently present / Total # of species present before

Sampling is another method used to measure biodiversity. For instance, to measure biodiversity in streams, sampling devices called kick screens are used (Fig. 4.1).

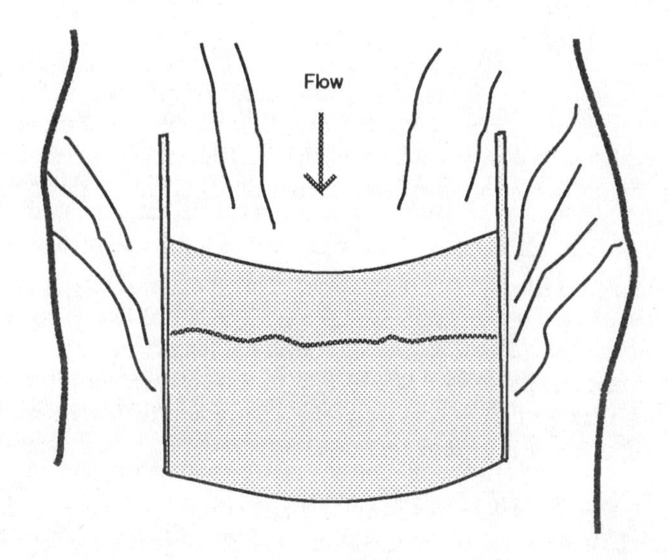

Figure 4.1 Kick screen

The screen is placed in the stream and used to collect small organisms. After examination under a microscope, organisms can be classified, and indices of species richness can be developed. One common index used is the EPT index, which counts the numbers of sample organisms belonging to three common orders Ephemeroptera (mayflies), Plecoptera (stoneflies), and Trichoptera (caddisflies). These orders contain many species which are sensitive to water quality changes. Generally, the higher the numbers of EPT taxa, the higher the water quality.

4.5 Ecosystem Patterns and Distributions

Concepts: spatial patterns, maps, cartographic scale

spatial patterns

The distribution of ecosystems is influenced by the location of natural features such as streams which control basic resources such as water, soil, and nutrients. The land area drained by a particular stream is called its drainage basin or watershed (Fig. 4.2).

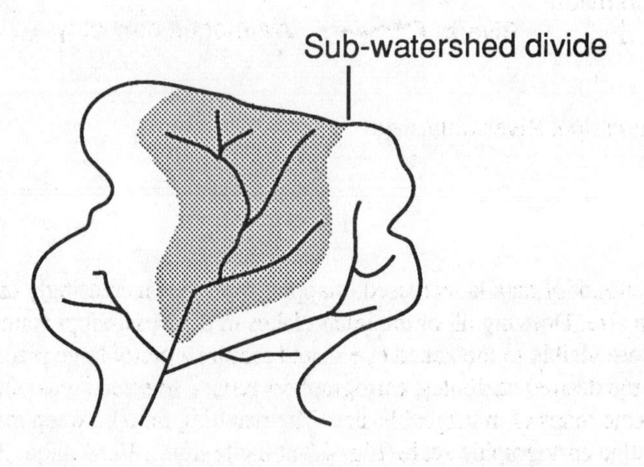

Figure 4.2 Watersheds

Characteristic patterns of stream size develop in watersheds. As streams flow toward the central and lower portions of the watershed, they grow larger as more and more water is added from sub-watersheds nested within larger watersheds. The sub-watershed divide is the topographic boundary that partitions runoff into different watersheds. The final product is a trunk stream, usually a major river, such as the Mississippi or Amazon (Fig. 4.3).

Suppose we want to see the small lakes in the Mississippi watershed. At this level of detail, we cannot tell where the small lakes are. We need another map, one of different scale. This is a good time to talk about maps and their use in environmental science.

maps

Maps are important tools for environmental analysis and understanding. They can be used to record processes such as floods, and to define patterns, linkages, and interrelationships at different scales. For instance, concentrations of inland lakes reveal the extent of glacial scouring in different areas. Maps can also show ecosystem disturbance. The use of satellite imagery for detecting flood damage from the 1993 Mississippi River flood is

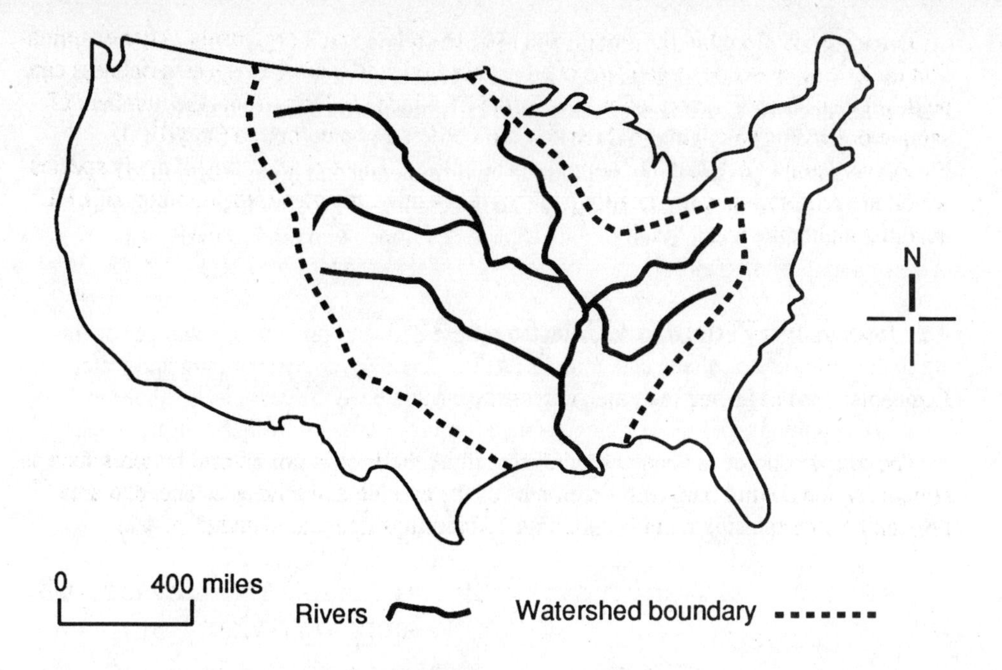

Figure 4.3 Mississippi River watershed

one example.

cartographic scale

When the amount of data is increased, mapping becomes increasingly difficult, given a constant map size. Drawing all of the inland lakes in the Mississippi watershed on a map so they were visible to the naked eye would require a pretty large piece of paper. In order to show the desired attributes, cartographers reduce selected areas of the earth's surface and create maps of manageable size. The resulting ratio between map size and earth size is called cartographic scale (Fig. 4.3 at the bottom). Here, about 3/4 of an inch on the map represents 400 miles on the earth's surface.

A large scale map is one where the ratio between map and earth size is relatively large, (e.g., one inch to one mile, or 1:63,360). Large scale maps show small areas, usually in great detail. Given an equal sized piece of paper, a smaller scale map would show more earth area (the denominator increases) within the same amount of map area; detail thus decreases. Examples of small scale maps include the world maps within this book.

Since maps show spatial relationships, maps can be used to help define questions for scientific investigation into the processes underlying the spatial forms. For instance, given a map showing the complex coastlines of Europe, one question is: What types of coastal erosion processes and engineering projects (such as the dikes and polders in the Netherlands), formed the coastlines?

Let's return to the distribution of ecosystems. The broad zonal pattern of ecosystems is broken by drainage systems and coastlines. These natural features alter the distribution of essential resources. River deltas and floodplains, such as the Nile provide examples of this. The river deposits soil from periodic floods and normal deposition, creating areas of high agricultural productivity in an otherwise hostile environment.

4.6 Models of Spatial Dynamics in Ecosystems

Concepts: succession, pioneer species, climax communities, disturbance

Ecosystems are constantly responding to environmental and human stimuli. These forces change the form, extent, pattern, and composition of ecosystems. A static map cannot capture these changes, so scientists have developed two models to describe them: succession and disturbance.

succession

Succession theory is used to describe the progressive and predictable changes occurring in communities of plants and animals as they spread over areas of land and water. Organisms at the early stages in the succession process have the capacity to alter the surface environment, and these changes make it more suitable for other communities. One community succeeds another, and eventually, a balance is established between one community and the bioclimatic condition of the biome. This relatively balanced, stable assemblage of organisms is called a climax community.

<u>In essence, species at the early levels of succession dig their own grave. How does this occur?</u>

pioneer species

Typically, succession begins with a barren landscape. Hardy organisms, called pioneer species, establish themselves in this tough environment, and then over time (many hundreds of years), stabilize the environment and pave the way for a larger and more complex community. Let's start at the beginning.

Lichens and mosses are pioneer species adapted to life on the rocks. As these species become attached to the rocks, they trap and accumulate airborne debris and some water. Lichens also produce acids which help decompose the rock. Eventually the breakdown of rock by these acids, coupled with the ongoing chemical and physical weathering by nature and the contribution of organic matter from dead lichens, creates a thin layer of soil.

This thin layer of soil creates conditions attractive to other organisms. New organisms such as worms, bacteria, insects, and other plants can now live in this environment. Since the lichen were better suited for the rocks, they lose out and are succeeded by this new community of organisms.

As the new organisms go through their life cycles, their deaths add organic matter to the soil, and this creates conditions suitable for grasses and other small plants. Larger animals can now find cover, and new communities develop. This succession continues until a complex community is established.

climax communities

The general trend in succession is for climax communities to exhibit more complexity and higher energy efficiency than the preceding successional communities. Climax communities have more organisms and complex interactions (mutualism, commensalism, predation, competition), than their successional predecessors. And, successional communities tend to gain energy by accumulating new matter, while climax communities are in energy balance, that is, the amounts of matter and energy entering and leaving the system are equivalent. Given this relative stability, climax communities can reproduce themselves; successional communities come and go.

The process above describes terrestrial succession, but it also applies to aquatic ecosystems. Here, ponds and lakes are gradually filled in by sediments from the surrounding land and the organic matter produced from the plants in the water. As more material accumulates, the shallower portions are able to support vegetation, such as cattails and sedges. The plant root systems and stems accumulate, and the water becomes shallower. Sediment continues to enter from the land, and soon the wet soil is dried, and grasses take over. Eventually the entire water area is filled in. Thousands of lakes and ponds in North America have met this fate. While most lakes are destined to decline, fill in, and become overgrown over hundreds, perhaps thousands of years, human activity, such as over-fertilization of nearby land, causes the entry of excess nutrients. This can shorten the life span of a lake.

disturbance

Opposed to this orderly view of ecological and progressive change, disturbance theory holds that constant change is the exception rather than the rule in ecosystems. Disturbance theory argues that ecosystems change more by external environmental forces such as fire, storms, drought, disease, and human pollution than by internal biological forces. The various disturbances acting individually and together create significant impacts on ecosystems and can alter their internal dynamics.

Disturbance theory also argues that these events operate at different intervals and magnitudes, and ecosystem change is a continuing response to the forces they unleash. Change comes irregularly with different magnitudes, as with hurricanes and floods. The environment is viewed as being more chaotic than under succession.

Which view is correct? They both are. Nature has long quiet periods of succession, and short chaotic periods of disturbance. Sometimes the disturbances are not so short, like the ice age. Human disturbances also impact the orderly procession of succession.

There are also locational differences between succession and disturbance. Some environments are more chaotic (dynamic) than others (e.g., seashores, mountain ranges, floodplains, and sand dunes). As agriculture pushes into these marginal areas, disturbances will increase.

4.7 Land Use as Ecosystems

Concepts: site and situation, scale

Human land-use systems are ecosystems with three characteristic traits. First, land-use systems require large amounts of energy from outside sources. Second, there is a progressive depletion of the resource base; and third, as human settlements encroach on marginal lands, there is increased risk from natural forces.

site and situation

The impacts of these negative characteristics on human land-use systems can be minimized if site and situation are used to proper advantage. Site refers to the actual location of the land-use system, and situation is a particular land-use system's linkages with the surrounding environment. Examples of favorable linkages for urban land-use systems include access to natural transportation (rivers, oceans), and proximity to productive agricultural areas.

We have already noted how marginal areas create gross mismatches between land uses and environment. A logical way to minimize impacts at the site is to choose activities well suited to the location. Avoiding dense development in floodplains is an example.

In terms of situation, using energy wisely is critical. If energy is wasted, multiple land-use systems may suffer. For instance, cities export their fossil fuel pollution (a non-renewable energy source) to other land uses (e.g., forests), in the form of acid precipitation.

scale

Site-specific land uses are quite precise in terms of geographic scale. For instance, aerial photography can accurately determine the extent of urban land use systems. However, scale relationships related to situation are more flexible. If we wanted to determine the effects of air pollution contributed by an urban land use, we would have to consider multiple scales. There are local impacts felt by residents near an emitting smokestack, but also distal effects if the contaminants contribute to acid precipitation and are carried downwind hundreds of miles. In addition, pollution emitted into the atmosphere may be carried by wind systems across many global pressure belts, thus requiring the consideration of this situation (linkage) at the global scale.

4.8 Human Impact and Disturbance of Ecosystems

Concepts: human impacts, matrices

human impacts

Environmental Geography defines six broad categories of human impact on ecosystems: reduction, fragmentation, substitution, simplification, contamination, and overgrowth. If we were interested in assessing the relative amount each of these impacts had on the environment, some kind of context would be required. For instance, stating that there is a 50% reduction of ecosystem area by human activity is meaningless without some context for comparison. What we really want to know is some part or all of the following: to what aspects of a specific ecosystem did the reduction occur; who or what caused the reduction; where and when the reduction occurred; and what happened? We can then begin to assess the relative impacts within some of these contexts.

To provide spatial context, eco-regions can be used. An eco-region is an area used for observation or classification; its boundaries are based on the dominant ecosystem present. Each eco-region may contain other smaller ecosystems.

Suppose we wish to assess the relative impacts on water-related ecosystems within an eco-region defined by a city, county, or province. To provide a timeframe, we can use the last five years, and we will use a city called City "X" of Exurbia as the hypothetical region. After studying current maps, three major types of eco-regions are identified: wetlands, streams, and lakes.

matrices

We now have a list of impacts (reduction, fragmentation, etc.), and a list of eco-regions. We can use a matrix to present this information. The following codes are used to represent the levels of each impact on the ecosystems: 0 = no amount, 1 = low amount, 2 = moderate amount, and 3 = high amount (Table 4-2).

The relative amount of each impact on different eco-regions is shown. The value of a "3" in a cell indicates that this impact was high on a specific eco-region. For instance, cell (1,1) shows the wetland area has been heavily reduced in City X.

Additional information is available from the column and row totals. The column totals tell us that contamination (with the highest total of 8), has had the highest overall impact on the different water related eco-regions in City X. The row total of 17 indicates that wetlands are the eco-region in City X receiving the most human impact.

TABLE 4-2 Relative Human Impacts on City X's Eco-Regions							
Eco-Region	Human Impacts						
	Red*	Frag	Subs	Simp	Cont	Over	Total
Wetlands	3	2	3	3	3	3	17
Streams	1	1	2	2	3	1	10
Lakes	0	0	1	1	2	0	4
Total	4	3	6	6	8	4	

* red=reduction, frag=fragmentation, subs=substitution, simp=simplification, cont=contamination, over=overgrowth

In real situations, it would be desirable to achieve a finer resolution of eco-regions, e.g., breaking wetlands down further into their lake-related (lacustrine), river related (riverine), and free-standing (palustrine) forms. Aerial photographs and maps from the National Wetland Inventory would be good places to start.

In order to do this, you would also need to justify your coding scheme, that is, use measurable criteria for what constitutes a low or high level of impact. An example of measurable criteria applied to reduction might look like this:

0 = no impact
1 = 10-20% loss of ecosystem area (low)
2 = 21-50% loss of ecosystem area (medium)
3 = greater than 50% area loss (high)

Coding is tricky and full of possible traps. Seek justification for your coding scheme from other scientific studies and field observations. Above all, be careful. If "high" impacts to wetlands are recognized by experts after a 20% loss, you might adjust your ranges accordingly. But you would have to be certain that this level of loss representing "high" applied to all the sub-categories of wetland you had defined.

Applications / Discussion

4.2) Record what you eat for one day. For each different food, trace the trail it follows from production, distribution, and to the market. You can get this information from product labels and calls to processing centers, mills, and retail grocers. For instance, if you eat a breakfast cereal, look at the ingredients on the box, such as rice or corn, and track their routes and mileage. Where was the rice grown? How far did it travel from its origin to the cereal mill? How far was the mill from the store where the cereal was purchased? The table below can help organize your work.

Food Chain Geography (example)				
Item	Ingredient	Origin	Processing Point (miles from origin)	Purchase Point (miles from origin)
cereal	corn	Iowa	Battle Creek, MI (400)	Boston, MA (900)
	sugar	Louisiana	Battle Creek, MI (1000)	Boston, MA (900)
	etc.			

In the example above, corn is grown in Iowa, and shipped to Battle Creek, Michigan 400 miles away. In Battle Creek the grain is processed, a corn cereal product is made, and then the cereal boxes are shipped to a retail grocery in Boston 900 miles away. Using the destinations above, this one ingredient has traveled 1300 miles.

Next, for the food you have eaten in one day, total up the mileage traveled by each ingredient. Is there a significant input of energy involved in the processing, packaging, and transportation of your food?

Lastly, for any item which can be called junk food, (e.g., potato chips, candy), compare the total ingredient mileage to a "healthy" food's mileage, such as vegetables. What type of foods have the higher mileages?

4.3) Using the data below (Table 4-1 from *Environmental Geography*), build a graph showing the relationship of species diversity and latitude (see example graph below).

Organism	Latitude			
	27°N	42°N	54°N	70°N
Beetles	4,000	2,000	164	90
Land snails	250	100	25	0
Reptiles	107	21	5	0
Amphibians	50	21	17	0
Flowering plants	2,500	1,650	390	218

Source: Clark, G.L. 1954. Elements of Ecology, Wiley, NY.

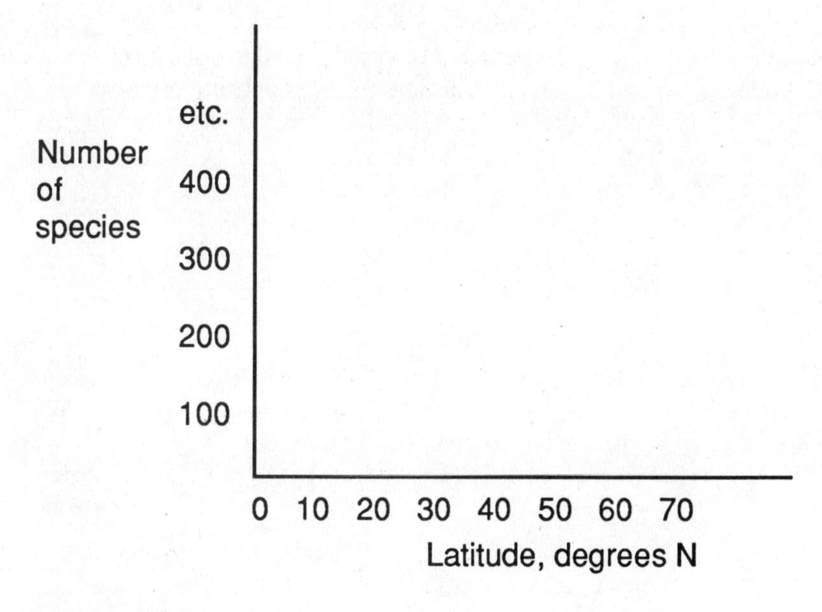

example graph for question 4.3

4.4) This application investigates the relationships between the wealth of nations and biodiversity loss. Create a table with four columns of data: 1) country; 2) per capita national income; 3) limitations of plant productivity related to climate; and, 4) major biodiversity loss.

Appendix A has the country list and the per capita national incomes. Use Figure 4.4 in *Environmental Geography* to obtain a general idea of the nations which have/do not have plant productivity limitations related to climate. If a nation has any part of its boundary within a non-limiting zone, code it as "0."

Your table should look something like this:

Country	Per capita national income	Limitations of plant productivity related to climate? 0=no, 1=yes	Major biodiversity loss? 0=no, 1=yes
Albania	$xxxx	1	0
Congo	$xxxx	0	1
.	.	.	.
.	.	.	.
.	.	.	.
Zimbabwe	$xxxx	1	0

Now calculate the average of the per capita incomes of all the countries that you have coded "0." Compare this to the average of the all the nations. Is the average per capita income of the countries without productivity limitations lower or higher than the total average?

Next, find these countries: United States, Japan, Germany, Canada, Great Britain, France, Switzerland, Sweden, and Italy. These are the most technologically advanced nations. What is their average per capita income? Were any of these nations coded a "0" in the limitations column?

As the last step, look at the map below and record the countries which have shown the greatest recent loss of biodiversity by entering a "1" in the column "biodiversity loss." Then calculate the average per capita incomes of these countries.

What patterns exist between the wealth of nations and biodiversity? What do these results suggest about the prospects and challenges for achieving political agreement on the use of technology within biodiverse regions?

4.5) Compare a map of world precipitation patterns and a physical map of the world (one which shows mountain ranges). For every continent (except Antarctica), what effects do the major mountain ranges (Himalayas, Rockies, Andes, Alps, Urals, Atlas, etc.) have on the regional distribution of precipitation?

4.6) What are the spatial and time-frame relationships between the climax communities of certain aquatic ecosystems (such as a pond), and the pioneer species of a terrestrial forest ecosystem?

4.7) Identify at least two of the site and situation advantages apparent in the major cities of New York, Los Angeles, Toronto, and Chicago.

4.8) What environmental impacts do animals have, e.g., beavers, birds, and house pets (cats and dogs)? Observe their behavior and make a list. How do animal activities impact human activities?

Recommended Reading

Wilson, E.O. (1992). *The Diversity of Life*. Harvard University Press, Cambridge MA.

World Resources Institute, United Nations Environmental Programme, and The World Conservation Union. (1992). *Global Biodiversity Strategy*. Washington, D.C.

Sources of Additional Information

INTERNET: BIOSCI, Biodiversity Gophers on the North American Internet Resources menu

Union of Concerned Scientists, 26 Church St., Cambridge, MA 02238, (617) 547-5552

World Resources Institute, 1709 New York Ave. NW, Suite 700, Washington, D.C. 20006, (202) 638-6300

References

Clark, G.L. (1954). *Elements of Ecology*. Wiley NY.

Robinson, A., R. Sale, and J. Morrison. (1978). *Elements of Cartography*. (4th. ed.), Wiley, New York, NY.

Rugg, D.S. (1972). *Spatial Foundations of Urbanism.* William C. Brown, Dubuque, IA.

Stapp, W., and M. Mitchell. (1991). *Field Manual for Water Quality Monitoring* (6th ed.). Thompson-Shore, Dexter, MI.

Wetzel, R.G. (1983). *Limnology* (2nd ed.). Saunders, Philadelphia. PA.

Whittaker, R.H. (1975). *Communities and Ecosystems* (2nd ed.). MacMillan, New York, NY.

5

THE HUMAN POPULATION: TRENDS AND PATTERNS

Review

5.1 Summary

During the period when agricultural societies dominated the world, population rose steadily, but slowly. When the revolutions in industry, transportation, and science occurred, these helped to jump-start the growth of the human population. Cities grew more populated as industry attracted workers from the fields. Large streams of migrants left their homes for new nations or new regions within their own country. It was as if something shook the planet and people fell into different locations.

Population dynamics and economic growth are closely related. Prior to industrialization, birth and death rates were relatively high. As industrialization and urbanization spread, the death rates dropped, and population grew rapidly in the more developed countries (MDCs). After many decades of urbanization, the birth rates in the MDCs began to fall due to the elimination of the need for family labor to support the farm. The increasing participation of women in society, birth control programs, and family planning continue to keep birth rates low in the MDCs.

In the less developed countries (LDCs), there are lower levels of industrialization, urbanization, and family planning than in the MDCs, but medical advances have lowered the death rates. Today, the gap between birth and death rates remains wide, and this maintains a high rate of natural increase. Since the majority of new births are in the already crowded LDCs, their ability to provide resources and space is shrinking.

It is the strain being placed on global and regional resources which concerns many people about the growing world population. Cultural practices often prevent the implementation of family planning, and wide differences still exist between nations on how to approach the population question.

5.2 A Historical Perspective of Population Trends

Concepts: trends, forecasts

Our concerns with population are based largely on the limited space and resources available on earth. Current consumption patterns also indicate that more resources are being used per person as living standards rise. For these reasons, trends in population distribution and consumption are viewed by many as important components of the population problem.

trends

Trends are general movements of a phenomenon over time in a certain direction. If we examine Figure 5.1, historical population growth reveals two trends: 1) Fundamental changes in agricultural and industrial productivity have led to general increases in the rate of population growth over the last 12,000 years; and, 2) The last 150 years has seen an over 500 percent (5-fold) increase in human population. Population quintupled; and then some. There were roughly 1 billion people on earth in 1850. Current population (1995) is 5.6 billion.

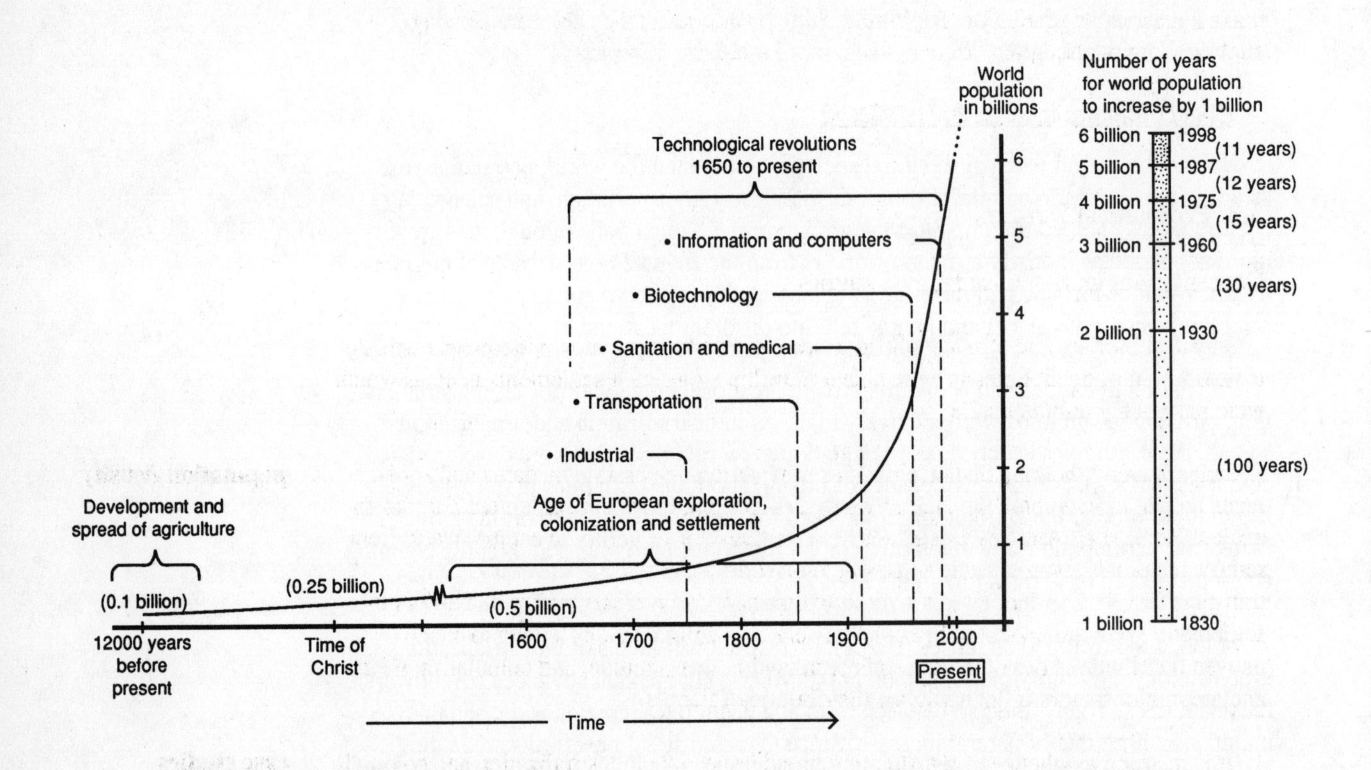

Figure 5.1 Wor!d population growth since the origin of agriculture

What growth pattern would you use from Figure 5.1 to predict population in the year 2000, or 2100?

Demographers use forecasts to try to predict future population. Simple population forecasting methods predict the future behavior of the population from its past trend. When there is more than one past trend, which trend do you use to predict what will happen next?

forecasts

Population growth is a complex process which involves a large number of variables; this complicates prediction. What if family planning efforts succeed in China and India by 1998? What if a major epidemic occurs, much like the one which wiped out one-third of Europe in the 1300's? These events are possible, but unlikely.

Predictions are based on what is most likely to occur. The key is to identify the factors which contribute most to the process, and then collect the most accurate data so you can make a reasonable prediction. Right now, high birth rates in the LDCs are slowing slightly, but not enough to "flatten" the curve for at least 30 years.

Could birthrates increase in the MDCs?

5.3 Population Patterns

Concepts: population density, case studies

Physical barriers and climate helped to form the early population patterns on earth. As technology improved, humans were able to develop significant settlements in areas which were previously uninhabitable.

Irrigation and the long distance transfer of water made possible by dams and aqueducts helped to overcome the lack of water as a limiting factor for settlement. Similar to ancient Rome, Los Angeles exists today only because of its ability to acquire water from remote locations. Even cities in relatively moist climates, such as New York City, transport their fresh water supply over long distances. As humans overcame barriers to settlement, great movements of people between and within nations helped to create uneven distributions of population. Industrialization, war, famine, and colonialism were, and are, major factors influencing the distribution of people.

population density

It is instructive whenever investigating broad issues like industrialization and colonialism to study a smaller area or "case" that helps to reveal many of the common issues or factors. With respect to population distribution, we can look at Europe as a case study of industrialization, and Nigeria as a case of colonialism.

case studies

Several geographic, historical, and cultural factors helped create a high population density throughout most of Europe. The continent benefited from the general distributions of land, minerals, and good soil, as well as its location relative to the other land masses.

In general, land and natural resources are more plentiful in the Northern Hemisphere. If you look at a world map, you can see that most land area is north of the equator. The world's major coal and petroleum reserves which played a key role in the industrial revolution are also located in the northern hemisphere. And, soils better suited for agriculture are found in the northern hemisphere. Europe, being in the northern hemisphere was able to cash in on these advantages.

An interesting perspective of Europe's relative locational advantage is shown by Figure 5.2.

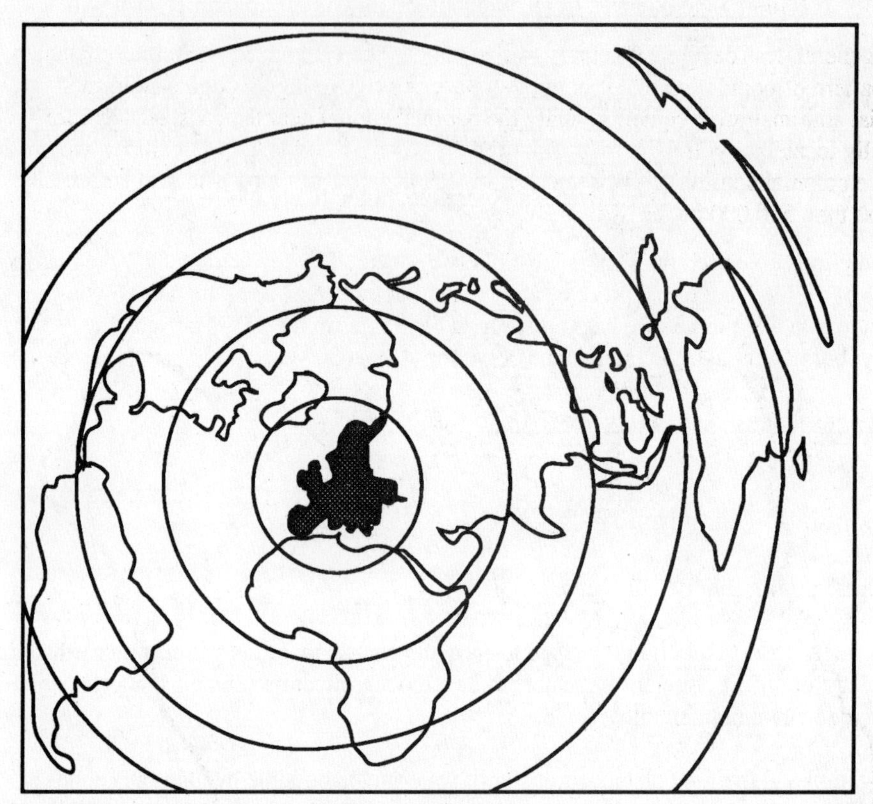

Figure 5.2 Europe's central location

Europe is centrally located with respect to the world's land masses. The idea that "all roads lead to Rome" was not just a saying, it was a fact. Europe also had coastal access from every direction, and no major physical barriers north of the Black Sea to prevent east/west exchange.

Timing also helped to spur European industrial development, and the subsequent population boom. Due to its central location, Europe was able to take advantage of other peoples' contributions, such as metal smelting from Africa and paper manufacturing from Egypt and China.

Progress was not strictly a European invention; at given times in history Indian, African, and Asian civilizations were more economically advanced than those in Europe. During these times the advances made were primarily related to agriculture (such as Mayan irrigation techniques). Industrial technology unleashed more powerful productive forces, and therefore resulted in greater population growth.

Let's look at the case of a former colony.

Nigeria was colonized by the British in the 1800's, and population was internally distributed by political and economic factors related to maintaining British colonial control. Principal among these factors were the establishment in Lagos of the administrative colonial center, and the imposition of a distorted economy which produced a limited number of goods for export to Britain, such as coffee and nuts. Local transportation networks evolved around the need to export local goods abroad to the colonizing nation.

In general, less developed nations—most having been colonized—exhibit a characteristic pattern of population distribution: there are one or two highly populated former colonial administrative centers, usually the national capital near the coast, and a more centrally located city. In Nigeria this pattern is shown by Lagos (pop. 2 million), which was the colonial administrative center, and Ibadan (pop. 1 million). The next largest city has less than 500,000 people (Fig. 5.3).

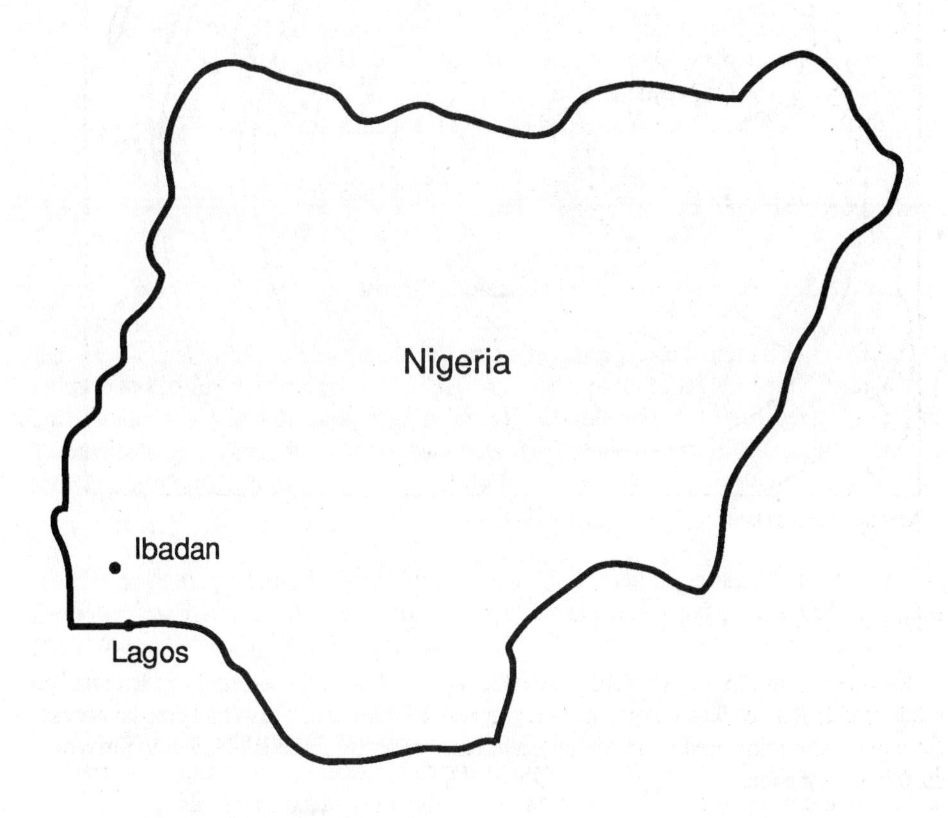

Figure 5.3 Nigerian urban centers

These examples represent a very brief look at case studies, but serve to illustrate the technique. One use of the results from case studies is to create theories. For instance, had we been the first to study population patterns under colonialism, we could have proposed the theory that former colonial administrative centers became the nuclei of population in those countries. The theory could have been tested by examining other cases, such as Senegal and Ghana. If significant differences between these cases and the original theory arose, the original theory might be updated with additional concepts, or replaced by another theory.

5.4 Changing Perceptions of Population Trends

Concept: perception

perception

Population problems are viewed differently by people from different cultures, but there is growing awareness around the world of population issues. This is reflected in the broad consensus reached at the 1994 United Nations International Conference on Population and Economic Development (ICPED).

Since 1974, at 10 year intervals, the United Nations has sponsored the ICPED. In 1974, the primary approach to population control was to promote contraception. This had the effect of splitting nations along religious and cultural lines, and as a result, full consensus was not achieved for curbing world population growth.

At the 1994 conference in Cairo, Egypt, a broader approach to controlling population growth was unanimously adopted by delegates from 175 countries. The Cairo plan sets no numerical goals for populations or birthrates. Instead, the plan asserts that population growth can be stabilized and enhanced by the advancement of women (Vig. 5.1).

Vignette 5.1 A New Approach to Population Growth

In LDCs, where population pressures are the greatest, there is widespread discrimination against women. According to UN figures, for example, girls are pulled out of school at younger ages than boys. Female workers are paid less than 40 percent as much as men for the same jobs. Girls and women are beaten more often than men, and women are typically excluded from political participation. And, many banks in LDCs routinely do not lend to women.

The new thinking argues that the old approach of simply promoting contraceptives was based on the false assumption this would limit family size. Instead, many women were alienated; feeling their economic, political, and cultural needs were treated as secondary to the sole motive for birth control.

The case of Kerala in southwestern India provides some results associated with the new thinking. Although one of India's poorest states, Kerala decided some years ago to initiate major campaigns to reduce infant mortality rates, educate all children, and provide at least small-scale bank credit to women.

Today, in Kerala, infant deaths are much rarer, with literacy almost 100 percent among boys and girls. And, banks make loans to women to finance such items as fertilizer or rain-tight tin roofs for houses. While India as a whole has a fertility rate of 3.7 births in a woman's lifetime, Kerala's rate has fallen to 2.0.

Education is seen as the key variable in this transformation, and this is supported by other data. Studies in 26 poor nations found that for every year of additional schooling for girls, especially beyond the third grade, the lower the desired family size, and the lower the actual family size.

In what ways do you think extending credit to women--even at a small scale--influences the population growth rate?

5.5 Natural Population Changes

Concept: public health and birth rates

Isn't natural population change based upon crude birth and death rates? Yes it is, but let's take a quick look behind the numbers.

Over 60 percent of the world's population does not have access to a clean supply of drinking water. In some LDCs, most families have lost a child or close relative due to a water-related disease. Diarrhea is a major cause of death among children in the LDCs.

public health and birth rates

Is there a connection between the perception of public health and the birth rate? If water quality in LDCs was improved so the deaths related to water-related diseases were cut in half, how might this affect birth rates?

5.6 The Demographic Transition

Concepts: demographic transition, LDC variant model

demographic transition

The demographic transition model illustrates the historical relationship between population dynamics and economic development (Fig. 5.4).

Prior to industrialization, birth and death rates are relatively high and roughly equivalent. In the second phase, while birth rates stay high, the death rate declines as medical care, public sanitation and diet improves. Near the end of the second phase a key effect of urbanization takes hold—less labor is required than on the farm. This effect continues, and along with the institution of family planning and birth control lowers the birth rate in the third phase. The final phase is characterized by low birth and death rates, and is a period of very slow, or zero population growth.

LDC variant model

The LDC variant model maintains the gap between birth and death rates. Death rates in LDCs remain low, but the birth rate stays high. Dispersion of the population, and the perception that children are economic assets for subsistence agriculture make family planning programs more difficult to implement. Future trends in population growth are uncertain because it is unclear what the effects of urbanization will be in the LDCs.

Under what circumstances might population growth in LDCs not decline, despite increases in the urbanization rates?

5.7 Fertility and Age Structure

Concepts: population pyramids, dependency ratio

population pyramids

Age-sex structures, or "population pyramids" allow you to obtain a quick overview of

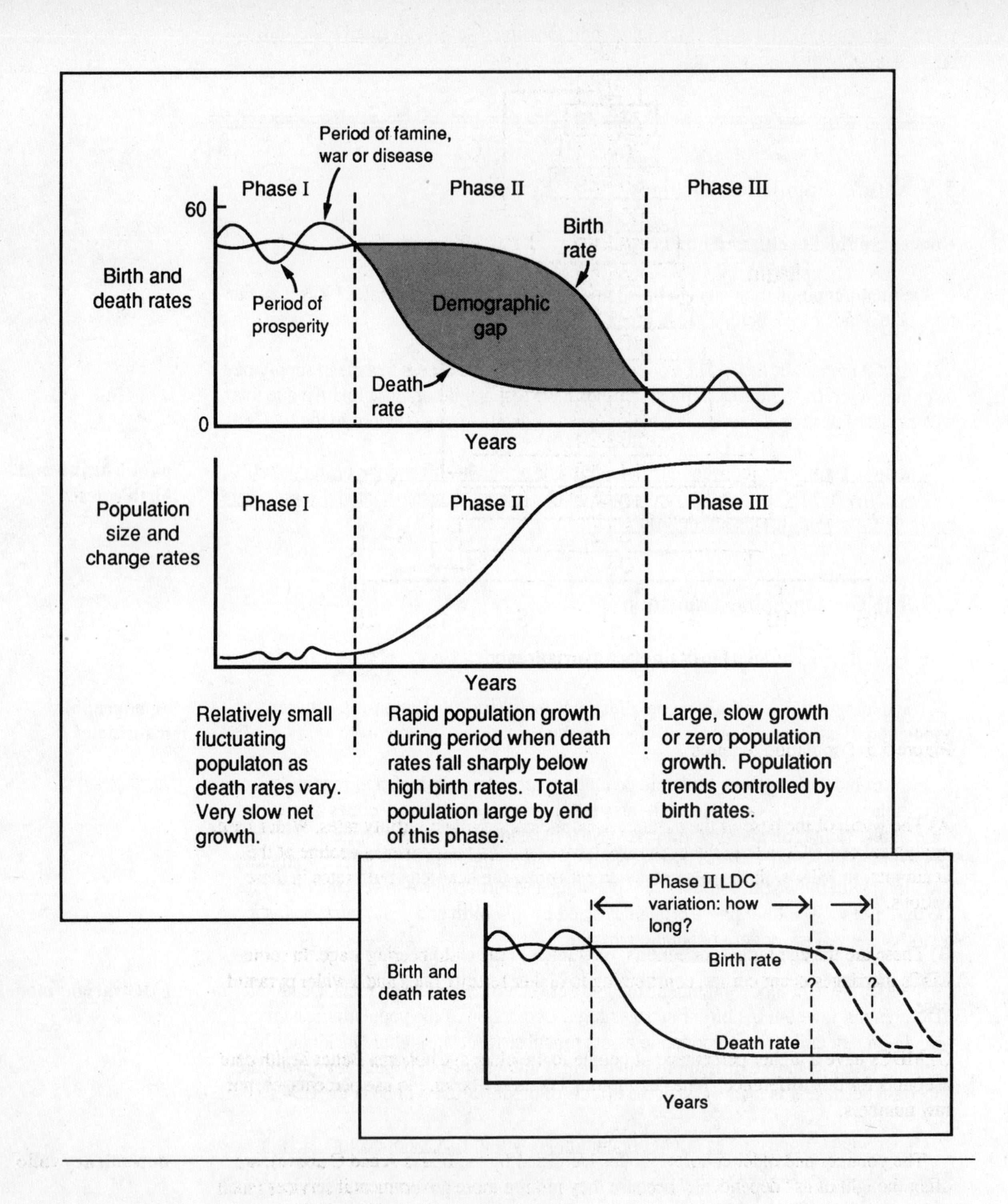

Figure 5.4 Demographic transition (with LDC variant)

population trends within a country. Figure 5.5 is a hypothetical population pyramid; the letters indicate some of the information available from this area of the diagram.

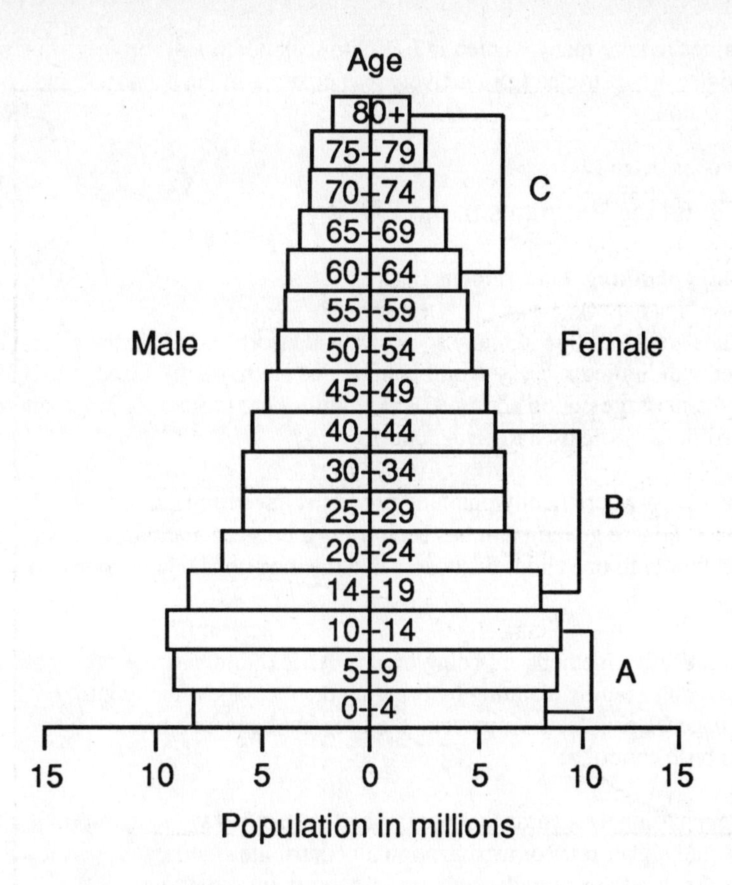

Figure 5.5 Population pyramid

A) The width of the base of the pyramid is an indication of the fertility rates. Wider bases are typical of LDCs, where the birth rates have remained high during a decline of the death rate. In MDCs, the base is narrower; reflecting the declining birth rates in these nations.

B) These are the age ranges, or cohorts, of women in the child bearing stage. In some LDCs marriages occur earlier, contributing to higher fertility rates and a wider pyramid base.

C) MDCs have a higher percentage of people in the older age cohorts. Better health care accounts for this difference. When comparing countries, be sure to use percentages, not raw numbers.

dependency ratio

The younger and older cohorts (groups identified by the letters A and C above), are often thought of as "dependent," because they require more governmental services (such as health care), than the middle age cohorts. One measure often cited here is the dependency ratio, which is obtained by dividing the total number of younger and older persons by the total population. LDCs and MDCs may have similar dependency ratios, but the main component of the ratio's numerator in LDCs would be younger persons, while in MDCs it would be the elderly.

It can be argued that the dependency ratio for LDCs should also include the women in the child bearing age cohorts. After giving birth, poor health care and the lack of employment opportunities require many women in LDCs to seek additional services. At the same time, many women are restricted from actively participating in the economic and political affairs of their nation.

5.8 Family Planning Programs in the LDCs

Concepts: family planning, land reform

Large families are a tradition in many societies, and efforts to reduce the birth rate in LDCs have met with limited success. Rural populations are more dispersed, and it is tougher to get the message out. In addition, large families are needed for labor, and social customs are traditionally opposed to birth control.

family planning

China is one nation where family planning programs have been successful in lowering the birthrate. The Chinese government has been using a mass educational program aimed at limiting each family to one child. Billboards and the mass media help spread the message.

Other communicative methods are being tried in Africa. Zimbabwe is taking an innovative approach to family planning by bringing the message directly into the dispersed rural villages. Using local volunteers, the program helps to educate women about health care and birth control.

There is an increasing awareness that upgrading the status of women, improving the rate of literacy, and higher rates of urbanization all contribute significantly to the success of these efforts. Land reform provides another area worth investigating.

land reform

Why land reform? As a legacy of colonialism, subsistence agriculture dominates the economies of many LDCs. Most people own very little, if any, land. The agricultural economies are characterized by intensive farming which supports the family, and as a result there is the requirement for labor and a desire for large families.

Aggregation of small farms into collectives may increase the efficiency of labor and reduce the incentive for larger families. However, this has the possibility of increasing the stream of migrants to urban areas—many of which are already overburdened with incoming migrants. Can an orderly transformation of agriculture—one which works to eliminate the incentive for more family labor—occur without generating major disruptions in urban areas?

5.9 A Worldwide Program to Stabilize World Population--A New Approach

Concept: linkages

linkages

Aspects of a new approach to population stabilization were presented in Vig. 5.1. Education and the empowerment of women are the key components of this new strategy. Quite often, there are additional benefits to society when basic problems such as illiteracy are overcome. Fundamental problems tend to have many linkages to other aspects of society. For instance, a higher literacy rate improves voter registration, and thus involves more people in the political affairs of their nation.

Can you think of any other positive side-effects that may result from the eradication of illiteracy?

5.10 Migration: Where Do People Go?

Concepts: distribution of migrants, measurement of migration

distribution of migrants

People migrate everywhere, and for many reasons. At the international level, political oppression, famine, and war have generated mass movements of people between nations. The African slave trade, the potato famine in Ireland, and people fleeing civil war in Rwanda are examples. Millions have also migrated voluntarily just trying to secure a better life.

Internal economic and political change can stimulate inter-regional migration within nations. The industrial boom of the early 1900's in the U.S. stimulated the movement of millions of people from the rural south to the urbanizing north, to cities such as Chicago, Detroit, and Cleveland.

Millions also move within their own metropolitan region. Residential mobility tends to reflect family life cycle changes. Age and education are the key indicators, as younger, more educated people tend to move more. Household changes in MDCs—the emergence of women into the workforce—has meant more than one wage earner per family. This sometimes hinders family mobility since the potential new location usually has to provide jobs for both wage earners. However, high divorce rates have kept the overall migration rates in MDCs constant.

measurement of migration

Direct measures of migration/mobility are actually difficult to obtain. Using the place of residence at a fixed prior date is best, however, the accuracy of this method varies significantly between nations. Some nations maintain population registers, which are the most accurate since they record the origin and destination of every move. Others, such as the United States, use a census (performed once every 10 years in the U.S.) to obtain the last residence from 5 years ago.

The question on the U.S. Census form is: "Did you live at the same address 5 years ago? If not, specify city, town, or county." While the answer gives you 2 fixed points at specific times, there are also several weaknesses with this approach: 1) people may have moved multiple times or moved out and back to the same address; this underestimates the number of movers; 2) people who moved within the 5 year interval but then died cannot answer the question; and, 3) you don't have data about respondents from the beginning of the period—e.g., their marital status, so you could not tell if married persons were more likely to move.

In LDCs, the direct measures of migration usually come by comparing the place of birth and current residence. This paints a general picture of the migration streams within a country, and often helps with the estimation of the rural to urban migration.

5.11 Problems Faced by LDCs with Rapidly Growing Populations

Concepts: poverty cycle, interdisciplinary approaches

poverty cycle

Environmental Geography identifies rapid population growth as a key component contributing to a cycle of poverty in many LDCs. The poverty cycle is a chain of events beginning with the inability of food production to keep pace with population growth, continues with conversion of marginal lands and land degradation, and culminates with unemployment and expanded urban poverty as agricultural workers migrate to cities seeking work. The cycle is sustained and exacerbated as population continues to grow.

The poverty cycle places a strain on MDC economies when resources are not conserved and food production lags; in LDCs these events can be catastrophic. On top of these problems, many of the poorer nations have distorted national economies—they have to rely on "cash crops" promoted during their colonial periods. The name "Banana Republics" reflects the one crop dominating the agriculture in some Central American nations.

There are basic environmental, social, and economic problems which need to be overcome in order to break the poverty cycle. There are no easy solutions. Let's brainstorm a little.

On the environmental front, water and soil have to be considered two of the key limiting factors of agricultural production--they require joint management. Too often, government-sponsored soil erosion programs neglect the impacts of eroded soils and the harmful contaminants they contain on water resources. Shortages of capital and technical personnel often prevent these efforts.

Socially, the diversity of cultural practices affects attitudes toward birth control. Abortion is outlawed by many nations, and family planning runs counter to the teachings of several religions, such as Catholicism and Islam. Innovations are likely to fail unless they are compatible with current social practices.

Economically, many nations might benefit by using regional development efforts similar to the European Economic Union (an alliance of nations to reduce economic barriers and promote cooperation in trade, labor, and pricing). These alliances may be able to provide low-tech sustainable methods for use by the developed nations in the areas of resource conservation. This expertise and experience could then be traded for needed commodities. This is another longshot.

Focusing separately on environmental issues, population control, or economic development will not work. Programs that sound easy to implement, such as birth control or family planning, are often very complex, and involve entire cultural histories. These examples begin to demonstrate the complexity of the poverty cycle; its solution will require a many-faceted and interdisciplinary approach.

5.12 Perceived Problems in MDCs with Stable or Declining Populations

Concept: environmental equity

There are positive and negative aspects to population stability. On the positive side, lower birth rates may create less pressure on the global resource base, which may also reduce the human impacts on the environment.

From a negative perspective, many policy makers fear that stable populations and slowing economic growth may lead to a decline of national prosperity. Remember, few politicians are willing to promise anything less than development, growth, jobs, and prosperity (all rising). Growth is often beneficial, but what is prosperity, and who is prosperous?

Environmental equity is part of the new environmental activism concerned with the fair distribution of resources, and remedying situations where environmental hazards, such as landfills and industrial pollution, are borne unequally by groups in society.

environmental equity

Consider the United States: the number of people below the poverty line is increasing, and many people (mostly poor and minorities) in our inner cities are being disproportionately exposed to polluted air, toxic soils, and lead.

Many now feel that real prosperity can only be achieved with a combination of sustainable resource use and environmental equity. The prosperous are those who seek these ends.

Applications / Discussion

5.2) Develop three different scenarios for world population growth in the next 20 years: a "high growth" scenario, where natural increase continues at its present rate of 1.6%; a "low growth" scenario where the rate of natural increase steadily declines from its current level of 1.6% to between 0.5 and 0.9 percent; and a "no growth" scenario, where the rate of natural increase declines rapidly and reaches zero.

To develop the high growth scenario, start with a 1995 population of 5.6 billion and calculate the yearly increase. For example 1996 = 0.016 * 5,600,000,000. For 1997, add the 1996 growth to 5.6 billion and calculate again, and so on. For the other two scenarios, also start with 5.6 billion people. The steadily declining growth scenario could use 0.016 * 5.6 billion for 1996, and 0.015 * (1996's total), and so on.

After you have performed the three sets of calculations, compare the final totals. Which scenario do you think is the most likely to occur? Why?

5.3) Using an atlas or almanac, calculate the gross density of the earth's population. This is the total number of people divided by the total land area, and the result is expressed as persons per square miles or square kilometers. Next, develop criteria for what constitutes inhabitable land, and then calculate the habitable (or net) density—for usable land only. Are the gross and net population densities significantly different? Would your calculations change significantly if you changed your criteria for what was habitable land?

5.4) Why does the education of women seem to have such a strong impact on reducing the birthrate in LDCs? One possibility is the relationship between literacy and the effectiveness of family planning efforts. Can you cite other reasons?

5.5) Between the airport and downtown Detroit, Michigan there is a huge tire along the interstate highway which keeps a running total of U.S. car production. Most people notice the size of the tire, and not the number of cars being produced. What would be an effective communication medium and context—using a running total—to get people's attention across the United States about rising world population?

5.6) Using the data below, compare the urban and rural population changes in MDCs and LDCs, including the forecasts for the year 2025. What do these data suggest for the LDC variant model of the demographic transition? A graph can help here. See example below.

Comparison of MDC / LDC Urbanization			
	Millions in urban/rural areas		
	1950	1980	2025
MDC's			
Urban	445	801	1192
Rural	386	334	204
% urban	54	71	85
LDC's			
Urban	289	973	3915
Rural	1382	2343	2865
% urban	17	29	58
Source: United Nations Population Yearbook, 1994.			

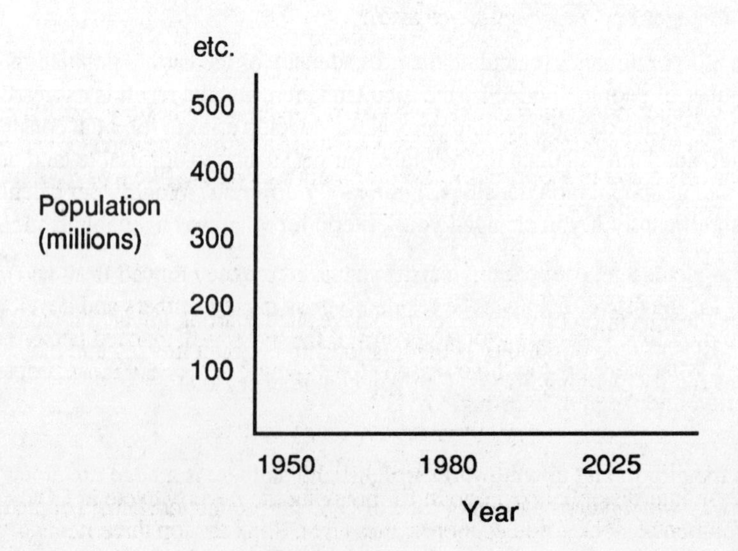

5.7) Use the data in Appendix A to compare the average ages of marriage and total fertility in three MDCs and three LDCs. Prepare a graph, and for each country plot the average age of women at marriage and the total fertility rate. See example below.

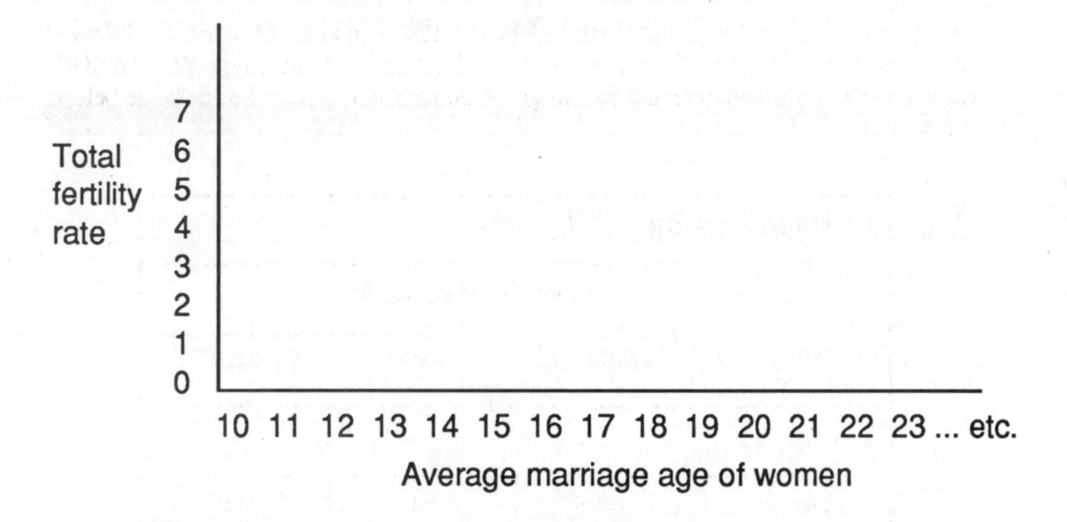

Discuss the similarities/differences between LDCs and MDCs.

5.8) Perform a case study of population control efforts in a LDC. You will need to consult some sources in the library. The United Nations has published many books on population case studies. What is the government's role? What agencies are involved? What is the participation rate? What cultural characteristics help or hinder family planning? Have the programs been successful?

5.9) Create a flowchart showing the positive side-effects from a reduction in illiteracy. Provide a description of the effects between each linkage. For example, if you show a link between the reduction of illiteracy and public health, your description could include how higher literacy rates might reduce sexually transmitted diseases. Do the side-effects have possible linkages between themselves also?

5.10) Interview 10 persons (non-students) who have moved in the last 5 years. To speed things up, you can place a public service ad in your student or local newspaper and have eligible respondents come to a central location for their interview—like a psych experiment.

Obtain respondents age, occupation, marital status, where they moved from and to, and the reason for the move. Tabulate the results from all class members and develop a profile of local mobility. What age group has moved the most? Are married persons more likely to move? What were the top three reasons for moving? Where did most people move—inter-city, inter-region, inter-nation?

5.11) Outline an interdisciplinary program for breaking the poverty cycle in LDCs. Propose environmental, social, and economic measures. Rank the top three items in each category in terms of importance for solving the problem.

5.12) What is urban sprawl, and how do land uses shift when it takes place? How are urban sprawl and environmental equity related? Are some land uses left behind? Are some people left behind? Discuss specific issues in the areas of air and water quality.

Recommended Reading

American Demographics (a periodical)

DeVanzo, J. (1982). *Techniques for Analysis of Migration History Data*. Publication N-1824-AID/NICHD. Rand Corporation, Santa Monica, CA.

Durning, A. (1990). "Ending Poverty," *State of the World 1990*. Worldwatch Institute, Washington, D.C.

Ehrlich, P.R. and A.H. Ehrlich. (1990). *The Population Explosion*. Doubleday, New York, NY.

Simon, J.L. (1981). *The Ultimate Resource*. Princeton Univ. Press, Princeton, NJ.

Sources of Additional Information

INTERNET: United Nations Gopher on the International Organizations Menu

United Nations. (1970). *Manual VI: Methods of Measuring Internal Migration*. Population Studies No. 47, New York, NY.

United Nations, *Demographic and Statistical Yearbooks*. New York, NY

References

Bogue, D. (1969). *The Principles of Demography*. Wiley, New York, NY.

De blij, H. and P. Muller. (1994). *Geography, Regions and Concepts* (7th ed.). Wiley, New York, NY.

United Nations. (1985). *Migration, Population Growth, and Employment in Metropolitan Areas of Selected Developing Countries*. Department of International Economic and Social Affairs ST/ESA/SER/R/57. New York, NY.

Todaro, M. (1980). "Internal Migration in Developing Countries," In Richard A. Easterlin (ed.), *Population and Economic Change in Developing Countries*. Univ. of Chicago Press, Chicago, IL.

6

AGRICULTURE, FOOD PRODUCTION, AND HUNGER

Review

6.1 Summary

Agriculture has been practiced for over 12,000 years. During this period, cultural contacts spread farming and animal domestication techniques to all locations on Earth outside the coldest regions. As agriculture paved the way for other forms of settlement, a sustainable balance (where the impacts were not severe enough to have caused major imbalances in other land-use systems) was initially achieved among early traditional societies.

The Industrial Revolution (1750-1850) spurred the specialization of labor in agriculture which was characterized by mechanization, larger farms, and higher yields. Environmental impacts also increased, as diverse ecosystems were altered or replaced, and external energy inputs in the forms of fertilizers and pesticides polluted the land and water.

Widespread hunger has resulted from inequitable distributions of food and poverty. Local adaptations to climate and terrain, and the history of political and economic relations between the more developed and less developed countries have helped to maintain the uneven distributions of agricultural capacity and poverty. Intensive subsistence farming dominates the less developed countries (LDCs); commercial farming is more prevalent in the more developed countries (MDCs).

Population growth, coupled with the degradation by subsistence and commercial farming of its basic resource base have raised concerns about the durability of agriculture. This has prompted the search for sustainable systems of agriculture which use renewable resources and minimize the impacts to the landscape.

6.2 Cultural Evolution and the Development of Global Agriculture

Concepts: adaptation, diffusion, sustainability

adaptation

Early agriculture represented a fundamental change in the relationship of people to nature. The domestication of plants and animals was an adaptation of nature which permitted sustainable outcomes. Seeds and animals could now be cultivated and bred to meet, and in many cases exceed, the needs of the people over many generations.

The combination of expanded productive capacity and favorable climatic conditions enabled the establishment of permanent, or sedentary agriculture. Stable settlements arose to handle the distribution of surplus agricultural products. Cities were formed, and this represented a new spatial and functional land use.

Cities created densities of population which were higher than agricultural settlements, and contained functions (services) performed by some parts of society that were not involved directly with food production. There was now a need to administer trade, laws, and promote culture. In a sense, agriculture had gone public.

diffusion

Local experience with the environment was socialized as different peoples came in contact with each other. Agriculture was being spread by diffusion—via the direct contact of people from different settlements practicing different methods of plant cultivation and animal domestication. Early city location was closely related to the agricultural and resource trade routes between different regions. For instance, ancient Troy in Asia Minor arose as an intermediate location between Asia and the Danube Valley in Europe, and around 3000 B.C. Byblos in Syria became a city because of Egypt's need for timber from Lebanon.

At this level of development, the interaction of agricultural and urban land use systems formed a sustainable cycle. Growth in agriculture and urban regions was complimentary to both. Figure 6.1 depicts this cycle.

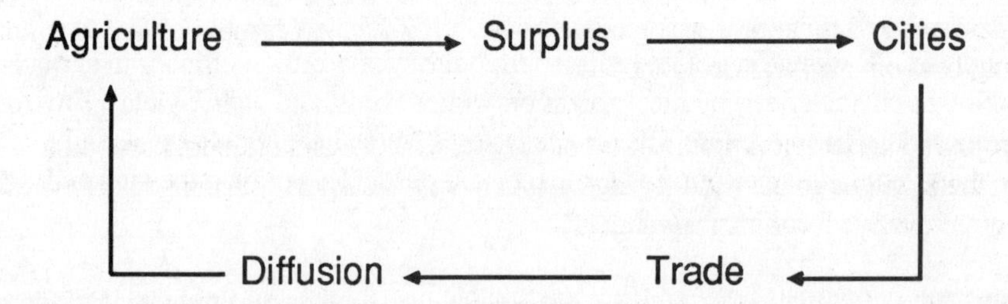

Figure 6.1 Sustainability of early agriculture

Agriculture created urban settlements through surplus production, and urban settlements helped to sustain agriculture as a source of demand for food. Cities became trade centers, which led to more frequent contact between peoples. New and complimentary products were exchanged between different regions. As trade centers, cities were the primary agents of agricultural diffusion.

sustainability

The development of early agriculture offers some insights into the relationship of sustainability, the current problems of world food production, and hunger. Sustainability can be achieved if: 1) there are enough goods produced to meet basic needs; and, 2) the methods used to achieve surplus production do not hinder the normal growth and operation of other land-use systems.

How would natural and cultural barriers influence the flow of events in Figure 6.1?

6.3 Industrialization and the Changing Nature of Agriculture

Concepts: mechanization, specialization, scale, distribution, sustainability

The industrial revolution transformed agriculture by replacing labor with machinery, and by establishing large-scale specialized crop and animal production. As population grew, market demand created new environmental adaptations. The new agriculture represented a simultaneous specialization of production methods and markets over larger scales.

mechanization

Mechanization is more efficient at larger scales. As scale in agriculture increases, the operating costs of machines per acre of cropland decreases. Specialization of crops reduces the "down time" incurred by switching machines, and many farm machines have diesel engines which can be used with greater efficiency on larger tracts because they are more powerful and require less maintenance than gasoline engines.

specialization

As agriculture adapted to markets, specialized regions developed for certain grains and livestock. Much of the sustainability for the early locations of regional agriculture was provided by proximity to transportation and related products.

scale

distribution

In the early part of this century, Omaha, Nebraska became a key stockyards location because of the proximal location of corn for livestock feed. The city's central location also gave it an economic advantage for shipping products via rail to distant population centers. However, advances in distribution, packaging, and production technology have undermined the initial relative locational advantages of regional agriculture. Since 1970, Omaha declined as a meat processing center when huge feedlots capable of fattening and processing beef on-site were developed.

sustainability

As crops and agricultural products are developed to withstand drought, pests, and spoilage, there will be impacts on regions which were, and are still sustained because of their original relative locational advantages.

Can you think of a case where technological advances, such as refrigeration, helped the early sites of regional agriculure? What characteristics of the product or site contributed to this?

6.4 Food Choices: The Plants and Animals that Feed the World

Concepts: yield, food value, limiting factors, storage

Five crops consisting of three cereals—wheat, corn, and rice, and two tubers—potatoes and manioc, dominate the world's food supply. Yield, food value, and storage are three characteristics shared by the world's successful staple crops.

These crops have evolved in different environments and are geographically distinct. Wheat is grown in the temperate and semi-arid regions of Europe, Central Asia, North America, South America, and Australia. Corn is found primarily in the humid temperate regions of North America, Europe, Africa, and South America. Rice dominates the tropics and subtropical areas of Asia. Potato production is centered in Eastern Europe and the old Soviet Union, and manioc is a tropical plant found in southeast Asia, Africa, and South America.

yield

Yield refers to the production per unit area of land. In subsistence farming, yields are closely related to inputs of labor; in commercial farming higher yields are obtained by inputs of outside energy in the form of pesticides, herbicides, fertilizers, and machinery.

food value

Food value refers to the caloric and nutrient levels of the crop. Since subsistence farmers must feed their families from crops grown on their own land, this is an important consideration. There are quantitative and qualitative aspects of food value, such as the relative amount of protein, and the percentage of protein usable by the body.

By weight, grains contain between 7 and 17 percent protein. Brown rice is at the low end of the scale, and triticale--a cross between rye and wheat, is at the high end. The protein content of one type of grain may vary significantly; wheat ranges between 9 and 14 percent protein. Hard red spring wheat can reach 14 percent protein, and the durum wheat used frequently in pasta has 13 percent, but these values are lowered when optimum moisture levels are exceeded.

The qualitative aspects of food value are related to the body's ability to use the available nutrients. With respect to protein, the food value of grains compared to meat is interesting.

By weight, meats contain more protein than grains, but our bodies are able to use as much or more of the protein from grains than from animal sources. Why?

limiting factors

The cause is related to the principle of limiting factors at work in nutrition.

Protein is made up of 22 amino acids. Eight of these 22 amino acids (essential amino acids or EAAs) cannot be synthesized by our own bodies—they must be supplied from outside sources. In addition, each of these EAAs must be present simultaneously and in certain proportions. Thus, the body can only use one pattern of these EAAs.

If one of the eight EAAs is below the required proportion in the pattern, the body synthesizes the amount of protein according to the lowest level provided. For instance, if a food contained 100 percent of a pattern's requirement for seven of the essential amino acids, but only 50 percent of the eighth EAA, then the body would use only 50 percent of the total protein. This is shown in Figure 6.2. The leftover amino acids are wasted as protein sources and treated as fuel in a way similar to carbohydrates.

When the usable protein between meat and grains are compared, it turns out that grains have equal or slightly higher protein efficiencies than meat. Rice has a 70 percent usability, the highest of the grains; the lowest is corn with just over 50 percent. Meat and poultry ranges between 65 and 70 percent usable protein.

Many proponents of vegetarian diets point out that adequate supplies of protein can be provided by grains, and we should end our reliance on meat as the primary source of protein. To grow one pound of beef, it takes 16 pounds of grain; 3 pounds of grain are needed for one pound of chicken. By substituting grain for meat, there would be a reduction in the amount of land devoted to growing grain to support the production of animal protein, and the difference could be used to help feed undernourished people.

storage

Finally, crops must have the ability to be stored. The top five crops have outer shells (rice, wheat), husks (corn), skins (potatoes), or exist as roots (manioc), which prevents their rapid decomposition. The characteristic pellet shape of the cereals is well suited for specialized storage and transfer areas, such as silos.

obtaining water.

The distribution of land is another key issue. In Mexico, landless peasants in the state of Chiapas have protested the failure of the government to implement land reforms and distribute land. The failure to do this hinders attempts to break the poverty cycle.

Globally, nations need to cooperate on the amounts and types of agricultural products produced. Is it possible to develop an equitable pricing structure which does not penalize those nations producing relatively smaller outputs of grain and other agricultural commodities? How could this be implemented?

6.7 Expanding Food Production: Some Pros and Cons of the Green Revolution

Concepts: green revolution, environmental dilemma, interrelationships

green revolution

The Green Revolution is the application of plant genetics to improve crop productivity. Hybrid strains of rice and wheat are used to boost grain production, and in many areas there have been substantial gains. However, there have been negative impacts associated with the capital and land intensive methods used to produce the Green Revolution crop varieties.

environmental dilemma

In some instances, the Green Revolution creates an environmental dilemma—that is, no matter what you do, you can't win. If you choose not to adopt improved crop strains, you are sacrificing potential gains in productivity. On the other hand, when you institute mechanized and capital intensive practices associated with "green" crops on small farms, these same farms are often inefficient, forcing many to merge and pushing tenant farmers off the land. Cities in many LDCs are already straining under the pressures of rural to urban migration. The United Nations estimates 20-30 million people migrate to cities yearly.

interrelationships

To achieve sustainability, there is a need to balance many different land use systems within a nation or region. In some parts of the MDCs there is a big push for replacing current management methods used in commercial agriculture with "sustainable agriculture." This is a positive step, but the conversion plans should also consider any related land use systems (urban, forest, range). In many rural areas, where small towns exist primarily to support local agriculture, a switch to sustainable practices is likely to change the flows of resources (e.g., labor, machinery, fertilizers) between the urban and rural land uses.

6.8 Outside the Green Revolution: New Strategies for Traditional Agriculture

Concept: linkages

For many LDCs, the question is not how an orderly transformation of agriculture occurs, it is whether a transition can be started at all. Limited capital and technical expertise limit agricultural development to those strategies which are demonstrably very efficient in their use of basic agricultural inputs: water, soil, labor, and energy (fertilizers, and pesticides).

linkages

One method for optimizing the range of agricultural development strategies is to look for approaches which address the major problems of resource use simultaneously.

Consider the land and water resources on a subsistence farm in a semi-arid region of an LDC. There is severe erosion from wind and water; water is scarce; and its quality is often degraded by human waste and parasites. A strategy benefiting both land and water would use local infiltration basins to temporarily capture rainwater and field runoff and recharge soil water and groundwater. Infiltration basins would reduce soil erosion by collecting runoff and sediment, and groundwater is usually superior in quality to surface water since it is generally more parasite-free.

Other linkages are possible. Can you think of a strategy for improving soil and fertilizer use simultaneously?

6.9 Beyond the Green Revolution: Biotechnology

Concept: genetic engineering

genetic engineering

Biotechnology is the use of scientific techniques for improving the productivity of crops and livestock. Genetic engineering is a new form of biotechnology which alters the genetic structure of plants and animals.

There is some controversy over the use of biotechnology, particularly its side effects on small farmers and animals. The recent approval of recombinant bovine growth hormone (rBGH, somatotropin) for use in milk has sparked much debate in these areas.

In terms of the possible effects on small farms, monitoring the rBGH program of injections requires sophisticated herd management techniques. Small farmers fear that widespread use of rBGH will favor large-scale factory farms. With respect to animal health, the drug works by triggering milk production in cows' mammary glands, and cows injected with the drug have shown a higher incidence of udder infections and welts.

6.10 Agriculture and the Environment

Concepts: human impacts, source control, law/science gap

human impacts

Agriculture has significant environmental impacts, including topsoil degradation, wind and water erosion, salt saturation, waterlogging, and groundwater depletion. The expansion of agriculture has also resulted in a reduction of wetlands, biodiversity, and forest cover. As population and the demand for food continue to grow, these impacts will continue, and perhaps become more severe as fields are pushed for higher yields and marginal lands are opened for new farming. There is a need to achieve a balance between expanding agricultural production and environmental preservation. What can be done from a regulatory perspective to cushion the blows to the landscape?

source control

Regulations are most effective when they address the source of the problem, rather than the effects. But many governmental policies miss this point. Taking the U.S. as an example, environmental regulation has focused on penalties after a problem or impact is detected, instead of focusing on reducing the problem at its source. Water use in the western U.S., which often results in over-irrigation and soil salinization, demonstrates the lack of source control.

In part due to the land grant history of U.S. agriculture, large farms (especially in the semi-arid west), have been subsidized to use water (this process began with the Reclamation Act of 1902). These subsidies provide water at a price from $3 to $15 per acre/foot—a price which offers no incentive to conserve. A positive feedback loop exists—the problem is increased by feedback. As more water is applied, salinization rises and there is a reduction of crop productivity.

Is western water really cheap? (Vig. 6.1).

Vignette 6.1 Western Water Subsidies

An acre-foot of water is 325,900 gallons. At current prices, a typical urban user of water (residential) pays $2.50 per 1000 gallons. This usually covers both supply and sewage treatment. At this rate, the residential user pays $812.50 for an acre-foot—the same amount of water sold to a large farm at $3-$15. Even if sewage treatment costs are taken out, the price is still over $400 per acre-foot, or 27 to 133 times greater than western farmers pay.

Pricing water at such a low rate creates no incentive to conserve, and we neglect one of the sources of the problem—excessive use.

Besides government programs, laws and regulations also miss the point.

law/science gap

Compounding the problem in the western U.S. are the surface water laws which discourage conservation, because acquisition of water rights (called appropriation rights) require continuous use of the water. In the West, you "use it or lose it," with respect to water rights.

Throughout the U.S., laws for controlling pollution also fail to reflect the natural processes at work. Groundwater is naturally connected in the hydrologic cycle to surface water, but most states in the U.S. do not have laws to protect groundwater (e.g., by limiting withdrawals) to complement their surface water laws.

Why do we retain such wasteful practices and the laws that encourage them? It is partially due to the influence of powerful ranching and agribusiness interests, and the failure to communicate scientific knowledge to policy makers. How can the exchange of information between scientists and policy makers in the federal and state governments be improved?

6.11 Moving Toward Sustainable Agricultural Production

Concept: sustainable agriculture

Concerns over the future of our food systems have prompted many to search for farming techniques that are not only profitable, but durable.

sustainable agriculture

Sustainable agriculture is a commitment to practices which achieve the permanence of soil, water, crops, and livestock. To reach sustainability, agriculture must substitute renewable resources generated on the farm for nonrenewable, purchased resources such as animal manure for commercial fertilizers. There must also be a switch to practices aimed at preserving the ecosystem and minimizing disturbance: crop rotation, landscape management, and livestock waste management.

A recent study by the Northwest Area Foundation (1994) compared sustainable farms to conventional farms in an eight state region of the United States (Minnesota, Iowa, North Dakota, South Dakota, Montana, Idaho, Washington, and Oregon). The key findings were:

1) Sustainable agriculture is a highly adaptable management technology which relies on intensive site-specific information, that is, improved knowledge of soils, water resources, and topography.

2) Sustainable agriculture has measurable environmental benefits, including reduced toxins in soil and water, less erosion, enhanced wildlife habitat, and lower energy use.

3) Sustainable agriculture can be economically competitive with conventional agriculture. However, to become the technology of choice for most farmers, sustainable technologies require further refinement.

4) Current public policies, especially federal commodity programs, discourage the adoption of sustainable agriculture.

5) Sustainable agriculture can provide new farming and business opportunities for people in rural communities, but local business infrastructure must respond to the different production and market needs of sustainable farmers.

6) Sustainable farms are more diversified in both crops and livestock than are conventional farms and have less land planted to the major commodities that, within current systems of production, deplete the soil most.

7) Sustainable farms use less commercial fertilizers, pesticides, and energy. Instead of these purchased inputs, they rotate crops, recycle plant nutrients and manure, and plant more soil-building crops than do conventional farmers. They also use more cover crops, strip crops, contour grass waterways, and field windbreaks to conserve moisture for crop production and reduce soil erosion.

8) It may be easier to start a farm with sustainable practices than to convert one that is heavily invested in conventional practices and technologies. Many beginning farmers may find sustainable agriculture attractive because it depends more on skilled labor and management and less on capital resources.

Point #5 above demonstrates the need to link different land use systems in order to achieve sustainability. This was a lesson of early sustainable agriculture—the complimentary balance between farms and their associated urban centers (see section 6.2). Point #8 has important implications for farms in LDCs. In LDCs, where intensive subsistence farming practices use heavy inputs of labor and low capital investment, the switch to sustainable farming is likely to be smoother.

6.12 Can Sustainable Agriculture Supply Future Global Demand?

Concept: the change to sustainable agriculture

the change to sustainable agriculture

The question in the heading above sparks reasons to be optimistic and pessimistic. From the optimistic perspective, we know that the basic inputs to agriculture—energy, soil, labor, and water are renewable. Sustainable methods are catching on in the MDCs, and the use of intensive subsistence farming in the LDCs is amenable to sustainable methods.

On the other hand, mismanagement of these resources limits their supply—a big reason for pessimism. Without political cooperation between nations, sustainable practices will be site-specific, and will lack the comprehensive application to larger scales involving related land uses.

Two key questions are: 1) <u>Will a majority of commercial farms in MDCs embrace sustainable methods; and, 2) As subsistence farming in LDCs makes the transition to commercial agriculture, will sustainable methods be adopted?</u>

<u>Applications / Discussion</u>

6.2) Within your county or province investigate the balance between agriculture and urban land uses. Can you find information on how much of the local agricultural product makes its way into local markets? How much is exported outside the region? What agricultural products are produced in the urban area(s) for use on local farms?

Clue: You might call the county agricultural extension agent, or the local office of the U.S. Department of Agriculture for estimates of local markets versus export.

6.3) For each state or province within your country, obtain the price of milk per gallon or liter and the milk production totals for each state or province. To obtain the prices, call the major supermarket chains. Compare the prices of milk in those states or provinces which are not large producers of milk to the prices of milk where milk production is concentrated. Has the distribution network eliminated locational advantage, or have other factors, such as government price supports eliminated regional price differentials?

6.4) Perform a protein analysis of your class. What foods do people get their protein from? Is the class "protein efficient?"

6.5) Evaluate local agriculture within the context of the six major human impacts on local ecosystems: reduction, fragmentation, substitution, simplification, contamination, and overgrowth. Focus on one kind of ecosystem such as stream valley corridors. Your instructor can provide you with maps or aerial photographs showing an area of rural land use.

6.6) Where is hunger found in your region? Is it rural or urban? Using income data from the 1990 U.S. Census data, prepare a map showing potential areas of low, medium, and high rates of hunger. What measures could be taken at the local scale to improve poor peoples' access to food?

6.7) Talk to the people at a local agricultural cooperative. What hybrid seeds are used by the farms in your region? What adaptations were made by the farmers to use these new plant varieties? How were yields affected?

6.8) Integrated pest management (IPM) is a technique which seeks to control pests by using biological controls in order to reduce pesticide use. How can crop selection and landscape management techniques improve pest management? For instance, are some crops resistant to certain pests? Cite specific examples.

6.9) Somebody states: "We have a surplus of milk in the United States, yet a new genetically-engineered drug was developed to increase milk production—this makes no sense." List several reasons why you agree or disagree with this statement.

6.10) Many states have developed Best Management Practices (BMPs) for agriculture. BMPs are recommended procedures for reducing human environmental disturbance. Obtain the agricultural BMPs for your state (either from the Departments of Natural Resources, Agriculture, or the Soil Conservation Service). Is there an emphasis on reducing pollutants at their sources? What are the strengths and weaknesses of the procedures? Do they reflect natural processes? Are they enforceable?

6.11) Obtain a copy of "A Better Row to Hoe" (available free from the address listed below) and discuss the key findings of the research results presented.

6.12) Create a scorecard (see example below) containing the optimistic and pessimistic factors for achieving sustainable agricultural production in the world. Do not weight the factors. Which side wins? Now weight some factors differently based on your assessment of their relative importance. Which side wins?

Sustainable Agriculture Scorecard		
Factors	Score (unweighted)	Score (weighted)
Optimistic		
renewable soil	5	12
genetic engineering	5	6
.	.	.
.	.	.
Totals		
Pessimistic		
water pollution	5	10
no cooperation between nations	5	15
.	.	.
.	.	.
Totals		

Recommended Reading

Ceres, publication of the Food and Agricultural Organization of the United Nations. New York, NY.

Faeth, P. (1994). "Building the Case for Sustainable Agriculture," *Environment* (36), 16-20, 34-39.

Makower, J. (1991). *The Green Consumer Supermarket Guide*. Penguin Books, New York, NY.

Rouse, W. (1991). "Who Decides About Biotech? The Clash Over Bovine Growth Hormone," *Technology Review* (94), 28-34.

Sources of Additional Information

Bread for the World, 802 Rhode Island Ave, N.E., Washington, D.C. 20018 (202) 269-0200

Northwest Area Foundation, Communications Department, 332 Minnesota Street, E-1201 St. Paul MN 55101-1373

INTERNET: Food and Agricultural Organization (FAO) menu on the United Nations Gopher

World Food Program (Rome, Italy) email: simong@wfp.org

United Nations, (1994). *UN Convention on Desertification, Information Program on Sustainable Development*. New York, NY.

United States Department of Agriculture, Economic Research Service, Beltsville, MD.

References

Curwen, E. C., and G. Hatt. (1953). *Plough and Pasture: the Early History of Farming*. Collier Books, New York, NY.

Grigg, D.B. (1974). *The Agricultural Systems of the World: An Environmental Approach.* Cambridge Univ. Press, London, England.

Lappe, F.M. (1975). *Diet for a Small Planet*. Ballantine Books, New York, NY.

Northwest Area Foundation. (1994). *A Better Row to Hoe*. St. Paul, MN.

Postal, S. (1985). *Conserving Water: The Untapped Alternative*, Worldwatch Paper #67. Worldwatch Institute, Washington, D.C.

Tarrant, J. (1990). "World Food Prospects for the 1990's," *Journal of Geography* (89), 234-238.

7

ENERGY GENERATION, USE, AND THE ENVIRONMENT

Review

7.1 Summary

Energy occupies a central role in ecosystems and human development. Ecosystems are based on energy and its transfer, and human development depends on energy-based technologies to provide food, clothing, and shelter. In these pursuits, humans have built their own ecosystems in the form of land use systems that support the earth's 5.6 billion people.

There is no energy shortage on earth. The sun provides enough energy for the planet's plants, animals, and humans many times over, and there are adequate reserves of fossil fuels for the next 100-200 years. But we are depleting the available sources of energy rapidly—coal, oil, and natural gas-- and there is a need to begin the conversion to safer, cleaner, and more reliable sources of energy.

These criteria for energy use provide all nations with challenges and opportunities, regardless of their level of development. In their desire to advance economically, less developed nations will have to resist the urge to implement energy projects at scales which may strain their economies and create large environmental disturbances. Developed nations need to identify and retrofit existing inefficiencies, and boost their research of fossil fuel alternatives. At the same time, all nations should adopt policies of long term energy sustainability and begin the transition to perpetual and renewable sources of energy.

7.2 Energy Resources and Technology

Concept: technology/energy interdependence

technology/energy interdependence

Energy and technology co-exist in their own mutual admiration society. Throughout history, energy has provided the foundation for technological advancement, and technology has returned the favor by supplying the means for using energy from different sources, and by enabling more sources of energy to be discovered.

It looks like energy owes one to technology.

Not really. Energy is the cause, and technology the effect of human labor. Think about it. If it hadn't been for the curvature in the earth, and the related decline in solar heating at higher latitudes, we might all be receiving similar amounts of solar energy, and our ancient ancestors might have been able to make it without fire.

Probably not, because humans needed to produce some energy to sustain life.

From another perspective, nighttime presents a real problem for our species—we can't see anything. Why? No direct solar energy reaches half the planet. So after dinner (which sustains the body's internal energy, or metabolism), the next thing ancient people did was go out and play with fire so they could see better. Torches were the first forms of lighting, and old oil lamps with wicks have been found in caves dating from the end of the Old Stone Age.

The industrial revolution provides additional examples of the interdependence between energy and technology. The improvement of the steam engine was based on the science of gases and on the development of the theory of heat. When Abraham Darby discovered how to make steel using coal instead of wood (actually, coke was used instead of charcoal), it allowed for a greater output of cheap steel; coal being more plentiful than wood at that time in England.

The demand for power to run machinery spurred us to look for new sources of energy. Before the industrial revolution, energy was primarily used to provide heat for warmth and light. Local supplies of wood and oil met these needs well until industry required greater energy production. We needed to transform and harness new sources of potential and kinetic energy to provide power for machines.

The problem was that we got off on the wrong foot.

Indirect sources of energy, such as coal and oil--the fossil fuels-- require additional inputs of energy to recover. Strike one against energy efficiency. Fossil fuels must also undergo a conversion (through burning) in order to release their energy as heat or electricity. As energy is converted into different forms, its ability to do work diminishes. Strike two.

We should not blame ourselves for getting behind in the count. Our knowledge of renewable sources of energy and energy efficiency were acquired after we started using fossil fuels. We have also learned the fuels currently supplying 80% of the world's energy are essentially finite and being depleted rapidly. And, pollution to our seas, groundwater, and air has resulted from the transport, storage, and use of fossil fuels.

Now we know. <u>How do we avoid striking out?</u>

7.3 The Global Energy Economy: Energy Sources and Uses

Concepts: renewable sources of energy, nonrenewable sources of energy, sustainability, energy efficiency

renewable sources of energy

Renewable energy sources are those capable of being formed or regenerated by natural processes. Biomass and water are renewable because they naturally replace and rehabilitate themselves. Vegetation is recreated through disturbance and succession, and the hydrologic cycle renews the global water supply.

Nonrenewable sources of energy are those with a slow rate of replacement, making them essentially finite. Deposits of fossil fuels, such as coal and oil, took millions of years to form, but we are using them at rates thousands of times faster than they are reforming.

nonrenewable sources of energy

sustainability

Sustainable resource use rests upon the balance of two rates: the rate at which resources are used, and the rate of replacement of those resources. The first rate—the rate at which resources are used—depends largely upon the efficiency of use.

energy efficiency

Application of the Second Law of Thermodynamics gives us the necessary insight into energy efficiency. Why not the First Law? The first law states that the amount of energy in the universe is constant—energy is neither created nor destroyed. When we use energy, we never destroy energy, we just change its form. What we really want to maximize or "conserve" about energy is its ability to do work. This is the value, or quality, of energy. As fossil fuels become depleted, it is important to know how much is wasted, or conversely, at what efficiencies they are being used.

An oil burning furnace used for heating rooms demonstrates the difference between First and Second Law efficiencies. The First Law states that the energy produced by burning the oil must be conserved as it flows from the furnace to the rooms. There will be some heat "loss" up the chimney, so the efficiency according to the First Law is the ratio of the amount of heat which is delivered to the rooms to the amount of heat generated when the oil is burned.

A typical efficiency for oil furnaces supplying warm air at 110° F. when the outside temperature is 32° F. is 60-65 percent. According to the First Law, the amount of fuel wasted is about 40 percent. In other words, if the process were 100 percent efficient, it would have been accomplished with 40 percent less oil. In practice, we can never reach 100 percent efficiency, since some energy is always dissipated as heat (lost) in the transfer.

Based on the First Law, oil furnaces look like a pretty good way to heat a home--but this reveals only part of the story.

The Second Law tells us that energy has no value unless it can be used to produce work as it flows from one place to another. Therefore, energy efficiency should be measured by how closely the amount of available work used to perform a task corresponds to the minimum amount of work required to do the task. Scientists call this type of efficiency "Second Law Efficiency," and it is the ratio of the least available work that could have done the job to the actual available work used to do the job.

To compute the Second Law efficiency for an oil furnace in our example you would need to compute the minimum amount of heat required to supply warm air at 110° F. while the outside temperature is 32° F. This is the numerator of the ratio. The denominator is obtained by computing the available work that is consumed when the oil is burned and the heat transferred to the rooms—this is the actual available work used for the job.

When the Second Law efficiency is calculated for the oil furnace, it turns out to be 8.2 percent. There is much more room for improvement than was suggested by the First Law. <u>Why is there such a big difference?</u>

The First Law focuses on the energy content of the fuel (the oil) and computes how much of it failed to reach a particular place (the rooms). This ignores alternative ways to heat the home and just evaluates a particular method. The Second Law efficiency focuses on the task and calculates how much work is needed to get it done; then it looks for the method for doing the job that uses the least amount of work possible. Why use a bulldozer to move a small pile of sand when you could use a shovel?

It turns out the best way to heat a home is with a heat pump, which has a Second Law efficiency of 20 percent. This is over twice the efficiency of the oil furnace. Table 7-1 shows Second Law efficiencies for major energy uses in the United States.

TABLE 7-1 Second Law Efficiencies for Different Tasks in the United States

Task	Second Law Efficiency (%)	Percent of National Energy Budget
Transportation	10	28
Space Heating, Water Heating, Cooking, Refrigeration, Lighting	3-6	36
Industry	25-30	36
Average	15	

Source: Commoner (1977).

In general, energy that is transferable at high temperatures is available for mechanical work, and is of relatively high quality. Since electricity is an efficient way to transform the energy of heat into the work needed for mechanical tasks, electric motors are efficient; this accounts for the higher efficiencies in industry.

Going back to space heating, when oil is burned in a furnace at 500° F. to heat a room to 70° F., high quality energy is being used for a low quality task. The same task could be accomplished by a low-quality energy source, such as the waste heat from a power plant. Only a small percentage of the buildings in the U.S. (primarily in New York) use waste heat from power plants.

From Table 7-1, it appears we could save 85 percent of the energy we now use. In practice, this cannot be achieved, but there are big gains possible in many areas. Sustainable energy practices must include matching energy sources with suitable tasks.

7.4 Fossil Fuels

Concept: sustainability

From the previous discussions, it would seem crazy to continue depending on fossil fuels—coal, oil, and natural gas—to supply most of our energy. Fossil fuels are non-renewable, possibly contribute to global warming, do contribute to acid rain, and are often wasted in their application. How did our dependence on these fuels emerge?

There are several reasons why: 1) fossil fuels are easily accessible in the earth's crust, and economical methods for their extraction have been developed; 2) fossil fuels are easy to use; controlled combustion exists at all scales—from automobiles to huge furnaces and steam turbines; 3) fossil fuels have a high energy content per unit mass; and, 4) they are easily transportable via pipelines, tankers, barges, or trains.

If you are having fun on one ride at a carnival, it is hard to get off and try another, even if it might give you more fun. The organization of the energy industry and the ownership patterns within the industry have enabled its members to have a lot of fun at the carnival, because they have made enormous profits by taking advantage of the four attributes of fossil fuels listed above. And, the absence of a national energy policy in the United States has made the ride bumpy for the population as a whole.

sustainability

For the total population, our "fun" depends on using energy most efficiently in order to guarantee a sustainable supply. The question we must answer is: should we continue expanding the production of fossil fuels to meet rising demand for the benefit of a few, or do we try to meet the demand more efficiently for the benefit of all? The answer is partially related to time—do we maximize short-term gain through profligate use, or attempt to achieve sustainability through long-term planning?

7.5 Nuclear Power

Concepts: cost/benefit analysis, risk

Nuclear power emerged as an alternative fuel largely on the basis that it was cheap—a very small amount of fissionable material can create large quantities of electricity. There are other advantages of nuclear power: compared to fossil fuel power plants, nuclear power generation is clean—there is no smoke, particulates, nitrogen, carbon dioxide, or sulfur emitted; there is less disturbance to the landscape from mining uranium than coal; transportation costs are less, since the quantities shipped are lower; and fissionable fuels can be reprocessed, so they may last indefinitely.

Then why do several countries (Switzerland, Germany, United States) have moratoriums on nuclear power development?

Coinciding with the advantages of nuclear power, there are several disadvantages: 1) radioactive wastes are difficult to dispose; 2) decommissioning or dismantling worn out plants are very expensive tasks; 3) nuclear weapons can be made from reprocessed plutonium; and, 4) nuclear power plants have had "accidents" which have killed many people, such as in Chernobyl.

cost/benefit analysis

Cost/benefit analysis is often used to determine if a project should be built or continued. With this method, all of the costs of a project (construction, maintenance, financing) are compared to the benefits (power supplied). If benefits exceed costs, then the project is usually identified as economically feasible.

One of the criticisms of traditional cost/benefit analysis is that the cost side of the balance sheet fails to include environmental damage. Many nuclear reactors in the U.S.

were built solely on the basis of a favorable ratio of economic benefits to economic costs. There was no consideration of their environmental effects—the environmental costs.

risk

The problem is that it is hard to measure or attach a dollar value to environmental damage. How much does it cost in dollars when good views are eliminated, or access to a beach is prevented? With nuclear power, what is the cost of the risk from nuclear power generation? How much does the potential or actual loss of life and injury equate to?

Within the current political climate, many nations have perceived these risks to be too high a price to pay for expanding their nuclear power generation capability, although some continue using it (Vig. 7.1).

Vignette 7.1 France's Nuclear Experience

France continues to rely on nuclear power for its energy needs. In 1994, the country obtained 80 percent of its electricity from this source, compared to 35 percent in Germany and 20 percent in the United States. New construction of nuclear facilities has been halted in Germany and the U.S.; but four nuclear plants are currently under construction in France.

After years without widespread controversy, the nuclear program is beginning to unravel. Technical problems, waste disposal headaches, a multi-billion dollar breeder reactor failure, and a huge ($30 billion) debt to the national treasury are contributing to an increasing concern over the future course of energy development. Construction of reactors slated for the late 1990's has now been postponed.

The opposition to the nuclear power trend in France includes not only prominent figures, such as Captain Jacques Cousteau, but other industrial companies that want to cogenerate their own electricity. The monopolistic structure of the French power industry has also drawn criticism from the European Union in Brussels which wants the market opened to independent suppliers of more economical electricity from other sources—such as gas-fired power plants and wind farms. Placing all of their eggs in one energy basket is beginning to haunt the French.

If you were an advisor to the French government, what would you recommend for their nuclear power policy--continue expansion, downsize, or remain at the same level?

7.6 Renewable and Perpetual Energy Resources

Concepts: decentralism, scale, solar energy

Renewable sources of energy are those replaced by nature, but their availability can be affected greatly by human disturbance. Biomass is a renewable source of fuel, but only if we use wood and other biomass at a rate which does not exceed their natural replacement.

Perpetual sources of energy are also replaced by nature, but their availability is largely independent of human modification. Solar energy, tidal power, and wind power vary in their global distribution, but they are essentially decentralized and result from forces beyond our control (solar radiation, gravity, and pressure cell differentials).

decentralism

Biomass is an important fuel for heating and cooking in LDCs. As these nations develop economically, there will be the need to find higher quality fuels for industry and transportation. Besides the efficiency benefits realized by matching the energy source with the task, the environmental impacts of scale must also be considered.

scale

Many LDCs such as China and Brazil have plans to build huge dams with major environmental consequences: dams may block fish migrations, sediments fill in the reservoir, and the floodplain of the river is submerged, which offsets the flood control capability of the dam. Given the environmental impacts and high capital costs of these projects, it seems more logical for LDCs to approach them at smaller scales.

solar energy

How can the LDCs obtain energy efficiency with limited capital investment and environmental impact? By using more from perpetual sources, such as the sun and wind. Since many LDCs are in areas of high insolation, using solar power is a logical alternative. Solar energy, delivered to the earth's surface at a relatively low temperature is a good source for low quality tasks, such as cooking or space heating (Fig 7.1).

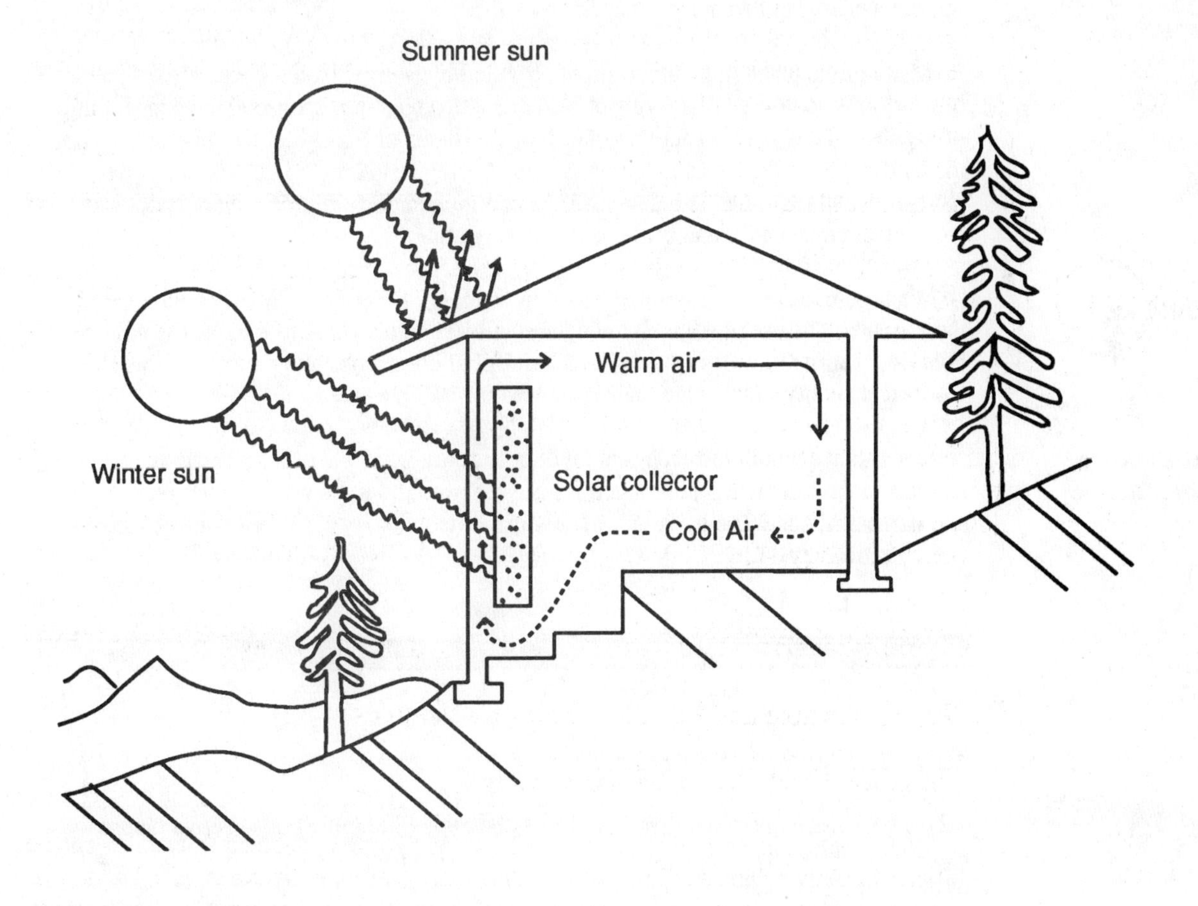

Figure 7.1 Passive solar heating

Solar energy is also a good source for high quality tasks, because the thermodynamic quality of radiant energy is determined by the temperature of the source emitting it. The surface temperature of the sun is over 10,000° F, and this higher quality energy can be obtained by concentrating solar energy with a mirror or reservoir. France has built a parabolic mirror in the Pyrenees Mountains which gathers enough light to produce a temperature of 6000° F. Hydropower is actually collected solar energy; diffuse rain (generated by solar energy) is concentrated in a reservoir and its falling motion is used to drive a turbine.

7.7 Using Energy More Efficiently

Concept: comprehensive energy efficiency

Our use of energy is currently out of balance. There is a need to optimize the match between energy sources and their applications, and to minimize the environmental impacts from energy use. These tasks are related, since wasted energy causes environmental disruption, such as the incomplete combustion of hydrocarbons and thermal pollution from power plants.

There are many immediate measures which can improve energy efficiency, such as replacing incandescent lighting with fluorescent lighting. In the long run, we have to make the commitment to use sustainable and perpetual supplies, and adjust energy use to society's changing demands.

Many of the technological advances provided by energy itself have increased the use of energy. When the spatial forms of settlement changed in response to the concentration of industry, there were no automobiles. Now, millions of fossil fuel burning engines dot the landscape, and cities have grown in size. Average trip length in automobiles has increased, and more fuel is being used. As communications improve, more people are working at home and urban settlements are spreading.

We need to adopt decentralized and renewable sources of energy to accommodate these trends. The use of solar and wind systems would also help to remove the stranglehold over energy supplies retained by a handful of companies and nations. Diffuse and renewable energy sources are logical choices for LDCs as well.

comprehensive energy efficiency

Our balance within the ecosystem will occur only when we minimize the thermodynamic and environmental contradictions arising from the ways we own energy, extract and harness it, distribute it, and ultimately use it (technology). This is comprehensive energy efficiency.

Applications / Discussion

7.2) The Alaskan pipeline traverses 1285 kilometers (800 miles) of sensitive tundra ecosystem. Various technologies has been applied to ensure the passage of oil, protection of the permafrost, and the protection of wildlife. Since the oil in the pipeline has a temperature of 35° C. (95° F.), much of the pipeline was built above ground so it would not thaw the permafrost. The buried sections of pipe are cooled with refrigerated brine.

To avoid disrupting wildlife migration patterns, several hundred crossings were constructed.

In California, a major consumptive use of energy is the California aqueduct, which transports water from the northern to southern part of the state. Is obtaining energy and resources worth all of the technology and energy we apply to get them? What can be done to minimize energy inputs to obtain energy?

7.3) Observe all of the activities and items which require energy in your home (cooking, water heating, appliances, etc.). Record the source of energy for each activity. Where can large improvements in First and Second law efficiencies be made?

7.4) Is it possible to achieve energy sustainability without long-term planning? Why or why not?

7.5) Since 1979 in the United States, an average of 50,000 persons per year have been killed by automobile accidents. There has never been a death due to an accident at a nuclear plant, yet no orders for new plants have been made since the 1979 accident at Three Mile Island. During this period, automobile ownership and use has grown. Why is there such a difference in perception between nuclear power and automobile safety?

7.6) Evaluate wind and solar power in terms of their ease of use, ease of access, energy contained per mass, and transportability. How do these forms of energy compare to fossil fuels in these respects?

7.7) Besides urban sprawl, what other technological and spatial changes are occurring in human land use systems (urban, agricultural) which might require modification of the ways we use energy?

Recommended Reading

Alternative Sources of Energy (periodical)

Brower, M. (1990). *Cool Energy: The Renewable Solution to Global Warming.* Union of Concerned Scientists, Cambridge MA.

Yergin, D. (1990). *The Prize: The Epic Quest for Oil, Money, and Power.* Simon and Schuster, NY.

Sources of Additional Information

Alliance to Save Energy, 1725 K Street NW Suite 914, Washington, D.C. 20006-1401. (202) 857-0666

Energy Probe/Probe International, 225 Brunswick Ave. Toronto, Ontario M5S 2M6

INTERNET: Energy and Resources Group, U.S. Department of Energy Gopher available on the North American Menu

World Resources Institute, 1709 New York Ave. NW, 7th Floor, Washington, D.C. 20006

Worldwatch Institute, 1776 Massachusetts Ave. NW. Washington, D.C. 20036. (202) 452-1999.

References

Commoner, B. (1977). *The Poverty of Power*. Bantam Books, New York, NY.

Flavin, C. (1994). "In France, the Nuclear Honeymoon is Over," *World Watch* (7), 6.

Fowler, J.M. (1984). *Energy and the Environment* (2nd ed.). McGraw-Hill, NY.

Golob, R. (1993). *The Almanac of Renewable Energy*. H. Holt, NY.

Melosi, M.V. (1985). *Coping with Abundance: Energy and Environment in Industrial America*. Temple Univ. Press, Philadelphia, PA.

8

THE ATMOSPHERIC ENVIRONMENT AND LAND USE

Review

8.1 Summary

Earth's atmosphere functions as a protective blanket and energy accountant. As a protective blanket, specific gases within the atmosphere absorb harmful radiation from the sun. From an accounting perspective, the atmosphere maintains the dynamic energy balance necessary for maintaining the conditions for life on earth. In order for the atmosphere to maintain its heat balance, the atmosphere and earth must reflect, absorb, scatter, and reradiate solar radiation.

Molecular density, air composition, and temperature vary with altitude in the atmosphere. This has produced four distinct layers—the troposphere, stratosphere, mesosphere and thermosphere. From an environmental perspective, the lowest layer—the troposphere, is the most important. It holds nearly all the water and oxygen, and generates the primary wind systems and weather.

The lower layer of the atmosphere is also most vulnerable to the effects of human activity on the landscape. Human activity in the last 200 years has caused substantial increases in carbon dioxide and methane, two gases which absorb heat. Scientists have identified a trend which indicates some warming of the atmosphere has occurred over the last 100 years. What is uncertain is whether the cause of the warming is a general flux in global temperature, or is due to the increased carbon dioxide and other "greenhouse" gases.

8.2 The General Composition of Air

Concepts: carbon dioxide, ozone, water vapor, phase changes, latent heat, scattering

Earth's atmosphere is a mechanical mixture of gases. Below 80 km (50 miles), the gases are relatively well mixed and in constant proportions throughout. This layer is known as the homosphere. Above 80 km, in the layer called the heterosphere, the gases are stratified according to their weights (Fig 8.1).

Nitrogen and oxygen occupy about 99 percent of the volume of the atmosphere. Most of the remaining 1 percent is comprised of argon, an inert gas of little importance to

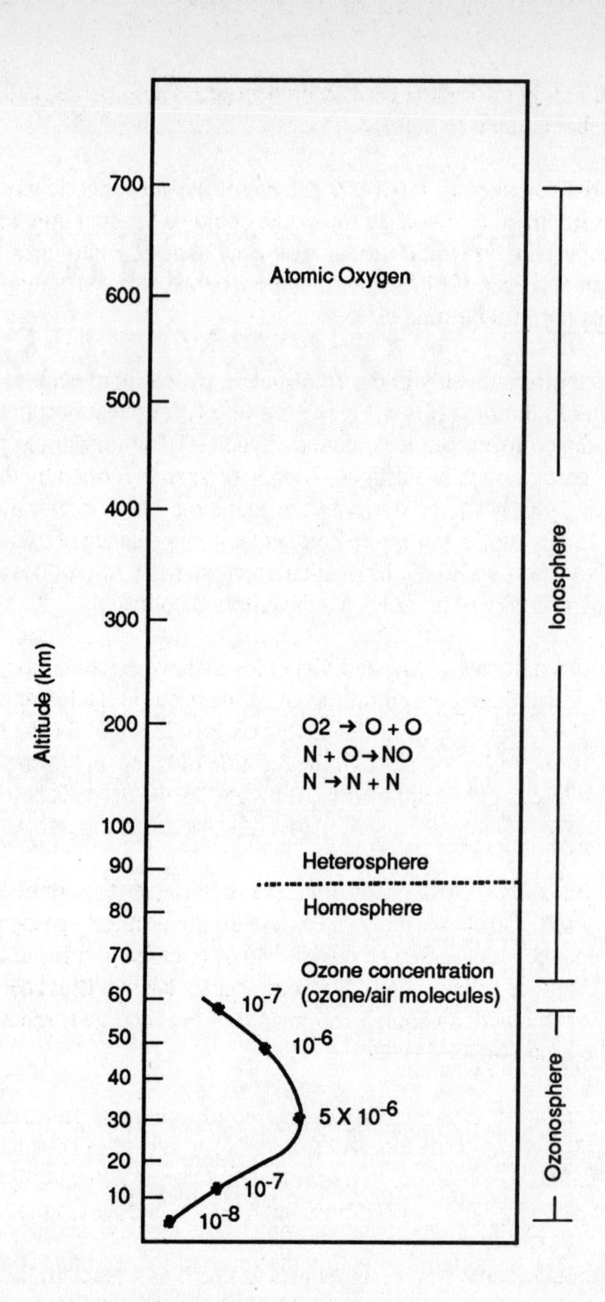

Figure 8.1 The main layers of atmospheric gases

natural processes. Oxygen is essential for respiration, but at sea level nitrogen is an inert gas as far as the body is concerned.

Carbon dioxide, ozone, water, and small particles are four minor constituents of the atmospheric mix with important roles. For instance, Carbon dioxide (CO_2) occurs naturally in the atmosphere and is an important absorber of infrared heat released from the earth's surface and lower atmosphere.

carbon dioxide

There has been a significant increase in atmospheric carbon dioxide since the industrial revolution of the early 1800's. This increase, resulting largely from the burning of wood and fossil fuels, has prompted concerns about human-induced climate change. Since CO_2 traps heat, there are concerns that rising levels may mimic a greenhouse and keep heat in, ultimately raising the earth's temperature. These trends, called global warming and global climate change, are discussed in section 8.8.

ozone

Ozone comprises an average 0.000007 percent of the atmospheric gas mix, but plays an important environmental role. If all the ozone could be brought down to sea level pressure and temperature, it would form a layer only about 2.5 mm thick (about one-tenth of an inch). As an absorber of ultraviolet radiation, ozone protects humans and other lifeforms from its harmful burning effects.

Ozone is distributed unevenly in the atmosphere, with a sharp peak occurring near the altitude of 25 km (15.5 miles) (Fig 8.1). The ozone (O_3) of the atmosphere is believed to be formed from the collision between atomic oxygen (O), molecular oxygen (O_2), and a third "catalytic" particle such as nitrogen. Atomic oxygen is formed by the splitting of molecular oxygen by very short waves of solar radiation. The concentration of the "ozone layer" between 15 km and 55 km is apparently due to the balance of two factors: the availability of very short wavelengths of solar energy needed to produce atomic oxygen; and, a high enough density of particles for the required collisions.

The distributions of ozone at low and higher levels have created important environmental concerns. Fairly high concentrations of ozone occur in the lower atmosphere, especially over urban areas. Much of the atomic oxygen required for the reaction described above is formed in smog through the reaction of smog and nitrogen dioxide, a by-product of automobile engine combustion. At higher levels, ozone is being depleted by chlorofluorocarbons—chemicals used in aerosol sprays and refrigerants.

water vapor

Another important constituent of the atmosphere is water. At normal earth temperatures, water is the only substance that can exist in all three states—gaseous, liquid, and solid. As water changes states, energy is absorbed or released. For instance, when water changes from a liquid or solid to a gas, water molecules take up heat energy from the air in which they are contained; an equivalent amount of heat energy is released when they revert back to the liquid or solid state.

phase changes

latent heat

These "phase changes" between the gaseous, solid, and liquid states of water play an important role in the energy transfers on the earth's surface. Heat consumed at one place during evaporation may be released at a new location during condensation. This "latent heat" is a source of energy for thunderstorms, tornadoes, and hurricanes.

Air always contains some water (sometimes as much as 4 percent), in the gaseous state in the form of water vapor. The atmosphere's capacity to take on water vapor is dependent on air temperature. Warmer air can hold more moisture, and the amount of atmospheric water vapor is related to the geography of land and water features. Deserts are so dry because they are fed with descending dry air from the upper atmosphere, and are usually isolated from oceans by distance or mountains.

All of the gases mentioned above and the remaining minor gases in the atmosphere are colorless. Why then, is the sky blue? (Vig. 8.1).

Vignette 8.1 Scattering

scattering

As solar radiation encounters very small particles in the atmosphere (gas molecules, dust, water droplets), the waves are deformed by these obstacles and dispersed in all directions. This energy dispersion is called scattering.

The nature of the scattering depends partly on the size of the particles, and the wavelength of the solar radiation. Gas molecules scatter at rates much higher in the short wavelengths (blue area of the electromagnetic spectrum) than at longer wavelengths (red area). This is why the sky appears blue—short waves are readily scattered by air molecules.

Other atmospheric colors are also related to scattering. At higher altitudes the sky becomes darker as air density and the amount of scattering decrease. The sky becomes black as the effects of scattering are eliminated. When the atmosphere contains large dust particles and water droplets (haze) from pollution and high humidity, short and long waves are scattered equally. The sky becomes less bluish, and milky in color. This is a common occurrence during hot weather pollution "dust domes" in major urban areas. A "clear blue" sky is a good sign that smoke, dust, and haze are absent.

What causes the colors of the sunset?

8.3 Thermal Structure of the Atmosphere

Concepts: pressure, ambient atmospheric lapse rate

Figure 8.2 shows the thermal structure of the atmosphere. At 110 miles altitude in the thermosphere, the temperature is pretty warm—about 1000° C. (1832° F.). Why then, in early 1995, did an astronaut aboard the United States' space shuttle complain of cold hands while taking a space walk at this altitude?

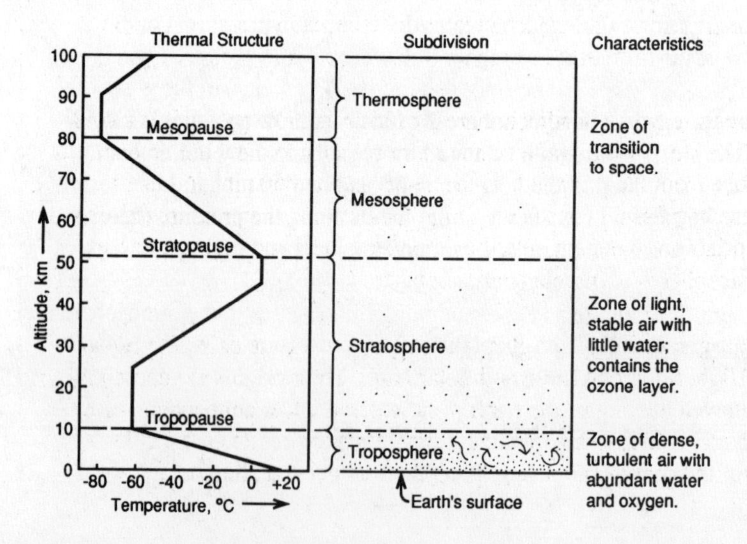

Figure 8.2 Thermal structure of the atmosphere

This area of the atmosphere is called the thermosphere, but only because individual particles have energies representative of high temperatures. At this altitude, there are very few gas particles, and because of this lower density, the total energy transmitted to a body through collisions is small compared to the energy of the "colder" particles at sea level. This is why the astronaut felt cold.

pressure

In general, as air rises it becomes less dense and loses pressure. When a gas changes pressure, its volume and/or temperature must also change. These relationships are called the ideal gas laws. You can feel this effect by letting air out of a bicycle tire; as the air expands and the pressure drops rapidly, the air cools.

ambient atmospheric lapse rate

Rising air in the atmosphere also loses density and pressure, and cools at a rate of -6.4° C. per 1000 m (-3.5° F. per 1000 ft.). This average rate of cooling in the troposphere is called the ambient atmospheric lapse rate.

In the stratosphere, temperature rises because of the absorption of ultraviolet radiation by ozone. The warming trend stops in the mesosphere, but is reversed again in the thermosphere, which is warm in a molecular sense, but feels cold to our bodies because of the low densities of air here.

From an environmental perspective, the troposphere is the most important atmospheric layer. All life and over 95 percent of the atmospheric air mass resides in the troposphere. Mixing within this layer is constant, and the primary wind systems and the majority of pollution are found here.

8.4 Atmospheric Pressure and General Circulation

Concepts: pressure, pressure belts, vertical air movement

pressure

Pressure is the force per unit area exerted on any plane surface. The atmosphere exerts pressure due to the weight of the air pressing down from above—from gravity. At sea level, the average atmospheric pressure is 1013.2 millibars (mb) or 14.7 pounds per square inch.

Uneven distributions of heat at the surface creates differences in pressure. On the earth's surface, the flow of air from areas of higher to lower pressure produces wind.

The difference in pressure between adjacent areas of high and low pressure is called the pressure gradient. The steeper the gradient, the more velocity to the wind. In some hurricanes, the difference from the perimeter to the center is often 50 mb, and this relatively sharp decrease in pressure causes very high winds along the pressure gradient. These pressure differentials are transient, and they represent local and regional energy transfers within the atmosphere's dynamic energy balance.

pressure belts

At a larger scale, four great belts of semipermanent global pressure exist: the Equatorial Low, Subtropical High, Subarctic Low, and Polar High. Their origins are caused by thermal origins—the uneven heating of the earth's surface and lower atmosphere, or by dynamic events—the convergence of two wind systems. *Environmental Geography* contains a description of the origins of each belt and their associated climatic and wind patterns.

vertical air movement

The vertical motion of air, although usually less than one-tenth of the horizontal speed, plays an important role in weather and atmospheric circulation. Vertical movement of air is primarily responsible for cloud formation and dissipation, as well as the transportation of particles to the upper atmosphere.

Once in the upper atmosphere, particles may move long distances across several horizontal belts of pressure. This process is enabled by the uneven distribution of ozone, causing temperature variations within the stratosphere. In turn, differences in pressure and strong seasonal wind circulations develop in this layer. The ozone hole over Antarctica, and pink tinges on glaciers in the Swiss Alps caused by wind-borne sands from the Sahara Desert are testaments to the particles being carried long distances.

8.5 Solar Radiation in the Atmosphere

Concepts: the sun, solar constant, waves

the sun

The sun is a pretty typical star; not very big as stars go, and not that hot either. But with a diameter of 1,400,000 km (870,000 mi.) and a thermonuclear engine, the sun generates a huge amount of heat (about 5.6×10^{27} calories every minute). The earth intercepts only about one part in 2 billion of this total.

solar constant

The average amount of solar radiation striking the earth's surface is called the solar constant. Measurements at the outer atmosphere indicate that the rate of energy delivery to the earth is 2.00 cal/cm^2/min (0.139 watts/cm^2). We are not certain about how long the output of the sun will continue, but given the size of its fuel supply, it will continue to shine for another 4 billion years or more.

Energy radiated from the sun behaves in part like waves and particles. The particle emanations are cosmic rays, and these are of little consequence on earth. The wave portion is the most significant, and this is how we will treat solar radiation.

waves

Waves are characterized by wavelengths, the distance between two successive wave crests or troughs), and frequency (the number of times a wave crest or trough passes a given point in a specified unit of time) (Fig. 8.3). Wavelengths are measured in micrometers (a millionth of a meter), and frequency is measured in hertz (Hz). A hertz is equal to one cycle per second.

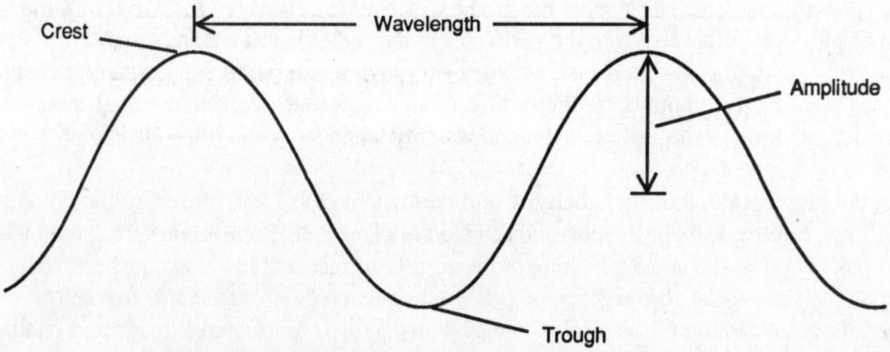

Figure 8.3 Characteristics of waves

8.6 Breakdown of the Solar Beam

Concepts: electromagnetic spectrum, albedo

electromagnetic spectrum

The waves emitted by the sun have different wavelengths. The range of wavelengths emitted by a body is called its spectrum; the sum total of all spectra is the electromagnetic spectrum. Figure 8.4 shows the spectrum for solar radiation. Most of the spectrum is invisible, with only the portion from 0.4 to 0.7 micrometers being visible.

According to Wien's Law, wavelength is a function of temperature—the hotter the radiating body, the shorter the wavelength of the emitted energy. Because of the sun's high temperature, most of the incoming solar radiation to earth is concentrated in the relatively short wavelengths of the electromagnetic spectrum. The cooler earth then re-radiates this shortwave energy at longer wavelengths.

Life on earth is enabled by the selective absorptivity of oxygen, ozone, water vapor, carbon dioxide and dust particles. Oxygen and ozone are the principal absorbers of harmful ultraviolet and other shortwave radiation; carbon dioxide and water vapor take care of radiation in the longwave portion of the spectrum (infrared).

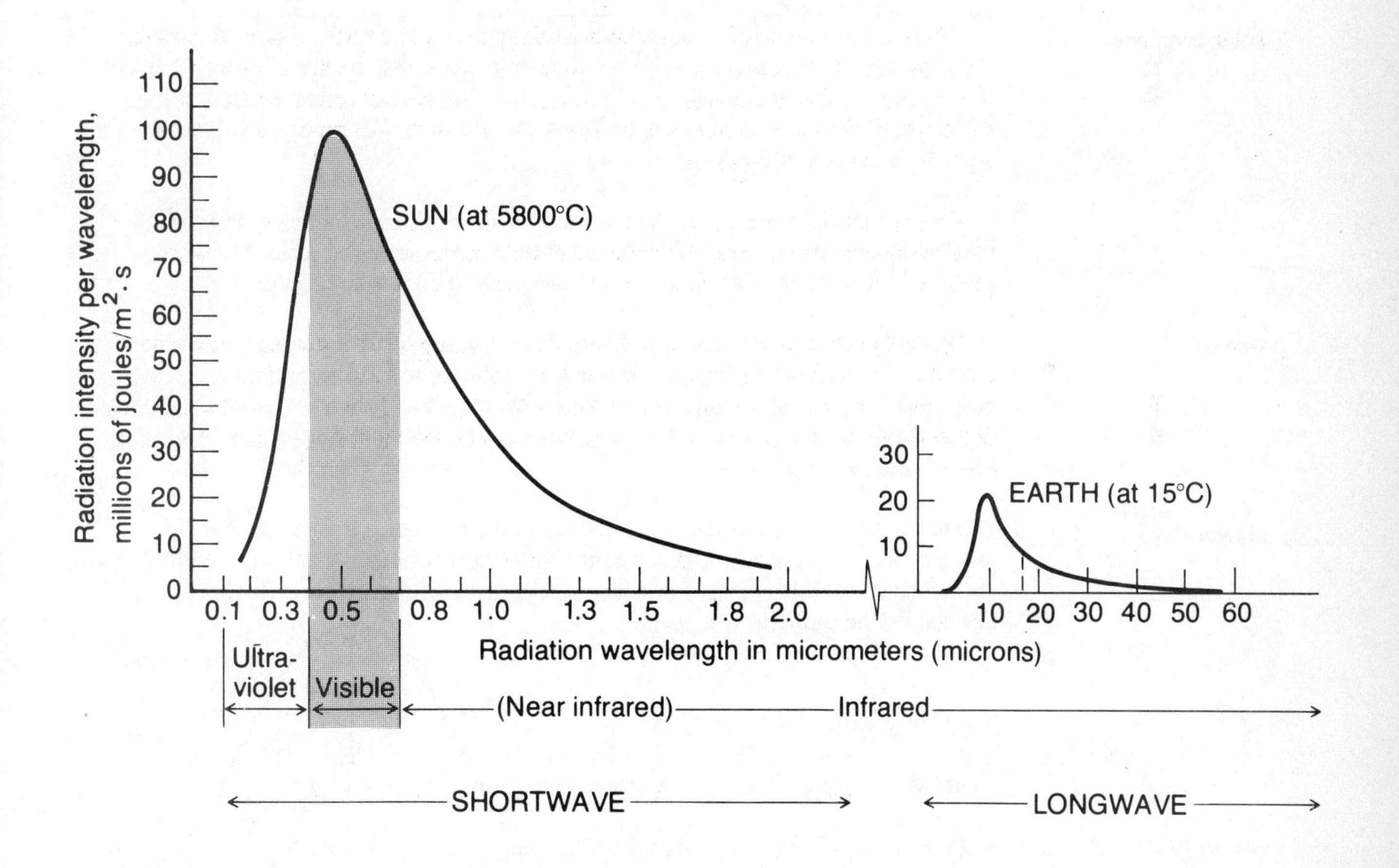

Figure 8.4 Breakdown of the solar beam

Figure 8.5 gives the relative sensitivity of the human eye to this visible portion. Maximum sensitivity is at 0.555 micrometers, in the green/yellow area. This helps to explain why emergency vehicles and school buses are painted these colors.

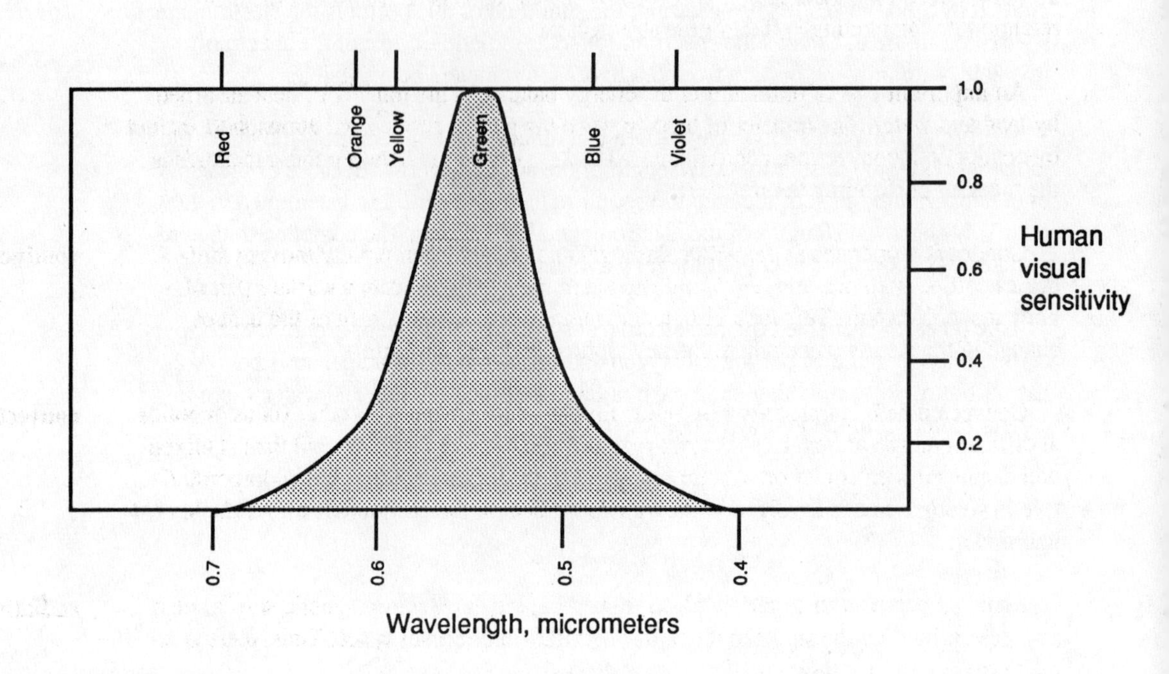

Figure 8.5 Relative sensitivity of the human eye

Overall, the amount of incoming solar radiation absorbed by the earth's surface is 46 percent; some as direct beams from the sun, some by cloud reflection, and the rest by atmospheric scattering. Adding this amount to the 22 percent absorbed in the atmosphere (9 percent shortwave and 13 percent longwave), yields a total of 68 percent. This is the amount of solar radiation which is absorbed by the atmosphere and earth's surface.

albedo

The remaining 32 percent of solar radiation is bounced off the earth. This is the average reflectivity, or albedo of the earth. For comparison, the moon's albedo is only about 7 percent. Of the planets; Mercury (0.07), Pluto (0.15), and Mars (0.30) have lower albedo than the earth, the rest are all brighter.

8.7 The Earth's Heat Balance

Concepts: energy balance, conduction, convection, radiation

energy balance

In section 3.3, the concept of the earth's energy balance was introduced. We know the earth has an equilibrium surface temperature of 15° C. (59° F), which reflects a long-term balance of incoming and outgoing heat.

The atmosphere is heated from above by solar radiation, and from below by the earth's surface. In order for the atmosphere to maintain its heat balance, the atmosphere and earth reflect, absorb, scatter, and reradiate solar radiation. Table 8-1 shows the balance of these components within the heat budget.

The percentages of energy emitted by the air to the earth (106), and by the earth to the air (114) exceed the total energy absorbed by the earth. This demonstrates the heat retention or "blanketing" effect of the atmosphere.

An important part of maintaining the energy balance is the transfer of heat absorbed by land and water. The transfer of heat between the earth's surface and atmosphere occurs by conduction, convection, and radiation. The key distinction between these methods is the medium performing the transfer.

conduction

Conduction operates at the molecular level and occurs when rapidly moving molecules collide with one another. Conducted heat always flows from a warmer part of a body to a colder part. This type of transfer is an important component of the heat exchange in the narrow zone where the atmosphere contacts the earth.

convection

Convected heat is transferred by fluids flowing into contact with other fluids or solids at different temperatures. Heat is lost or gained at the contact surface, and then is mixed and displaced with cooler or warmer parts of the fluid. Convection plays an important role in weather, and is involved in the transfer of heat in the atmosphere above the lowest half meter.

radiation

Radiated heat is transferred by electromagnetic waves. Electromagnetic waves, such as those emitted by the sun, can move through the void of outer space. Thus, there is no medium required for their transfer.

Boiling water in a pan illustrates the three heat transfer mechanisms simultaneously. Heat from the burner is transferred through the pan by conduction. This warms the lower layer of water in the pan, and the rest of the water above this lower layer is heated by convection currents which carry the water throughout the pan. You can see this when water boils. If you stand near the stove, radiated heat from the pan and burner reaches you, but with little effect on the intervening air.

8.8 The Greenhouse Effect and Global Warming

Concepts: greenhouse effect, causation

greenhouse effect

After striking the earth, the shortwave solar radiation is reradiated at longer wavelengths from the cooler earth in the infrared portion of the spectrum. Atmospheric gases such as carbon dioxide and water vapor do not absorb large quantities of incoming shortwave solar radiation, but they are good absorbers of longwave radiation. When a higher amount of the gases which absorb longwave radiation are present, more heat is retained in the atmosphere. This phenomenon is called the greenhouse effect, and the heat absorbing gases are the greenhouse gases.

causation

Human activity in the last 200 years has caused substantial increases in carbon dioxide and methane, two of the greenhouse gases. Scientists have identified a trend which indicates some warming of the atmosphere has occurred over the last 100 years. What is uncertain is whether the cause of the warming is a general flux in global temperature, or is due to the increased carbon dioxide and other greenhouse gases.

TABLE 8-1 Energy Budget of the Atmosphere (as percentages)

Incoming (shortwave solar radiation)	100
Absorbed by:	
atmosphere and clouds	22
earth's surface	46
Net shortwave absorbed	68
Outgoing shortwave solar radiation lost to space by:	
reflection from clouds	26
reflection from earth	2
scattering from air	4
Net shortwave outgoing	32 (a)
Additional outgoing radiation from the atmosphere and the earth	
The atmosphere:	
radiates to earth	106
radiates to space	60
receives from earth via radiation	-114
receives from earth via convection, turbulence, evaporation	-32
Net outgoing from atmosphere	20 (b)
The earth:	
radiates to space	8
radiates to the atmosphere	114
loses to the atmosphere via convection and turbulence	8
loses to the atmosphere via evaporation	24
receives from the atmosphere through radiation	-106
Net outgoing from earth	48 (c)
Total Net Outgoing (a + b + c)	100

Source: Miller and Thompson (1975)

As the glacial periods demonstrate, short and long term climatic trends are not unique occurrences in earth history. However, it is the timing aspect of causation which has scientists concerned about the greenhouse effect. Not only has warming followed the increase in the amount of carbon dioxide, but the increase has occurred over a short time span. Many scientists feel that the offsetting mechanisms may not have enough time to balance the increased energy absorbed by the greenhouse gases.

8.9 Some Consequences of Atmospheric Warming

Concept: models

Atmospheric warming involves many systems with a complex array of potential feedback effects. For instance, temperature increases might trigger secondary changes in the semipermanent pressure systems, global circulation patterns, and precipitation patterns. Or, melting of the polar ice caps and glaciers may occur; this could raise the level of the oceans and adversely affect millions of people in coastal areas.

models

To minimize complexity, scientists use models to simulate the effects of atmospheric warming. In general, models try to identify the major elements of a system and describe the strengths and direction of the linkages between those elements. A good model of a system also tries to explain the processes underlying the elements and linkages.

Review of a simple model of atmospheric warming demonstrates the complexity involved with this process and its effects (Figure 8.6.).

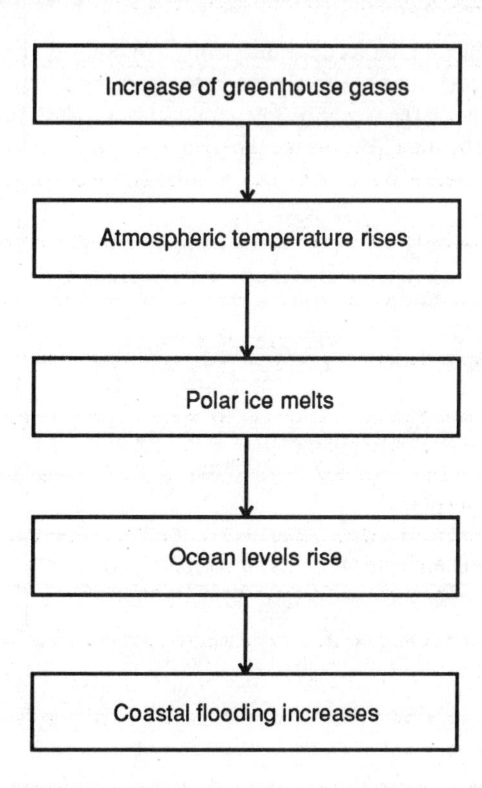

Figure 8.6 Simple climate change model

The process begins with an increase of greenhouse gases, and follows a sequence of events culminating in rising ocean levels and increased flooding. This model is simple, which is a desirable property, but it is also inadequate. Why? Because it fails to identify the processes underlying the increase of greenhouse gas. And, there are numerous other omissions: How might human responses to the intermediate events affect the outcome? Would rising temperatures cause an increase in ocean evaporation, and offset the rise in water levels? What are the effects of increased coastal flooding?

One of the problems with climate change models is their own complexity; this has hampered our ability to draw firm conclusions about atmospheric warming.

8.10 Urban Climate

Concepts: heat islands, environmental hazards

heat islands

Urban development retains and concentrates heat energy produced by the sun and human activity. Heat absorbing materials, such as concrete and tar, replace less absorbent grass and open soil. Homes, buildings, and factories emit excess heat and particles into the atmosphere. The result is a heat island—a zone or zones within cities with a mix of warmer, more polluted air than the adjacent urban or rural areas.

environmental hazards

The increased heat and particulate matter in the heat island increases convectional rainfall during the summer. Storms tend to be more intense with larger volumes of runoff. This provides a good example of how an environmental hazard can result from the interaction of human activity (urbanization) and natural systems (precipitation).

Minor differences in precipitation can significantly affect the potential for urban floods. The same pavement which absorbs more heat also blocks the infiltration of stormwater into the ground. Excess runoff is generated, which finds its way quickly through pipes into local streams. Streams receive a huge flush of water, and flooding often occurs.

How much water are we talking about? Over 1 square mile, an increase in precipitation of a hundredth of an inch delivers this much more water to the surface:

$$640 \text{ acres/mi}^2 * 325{,}900 \text{ gal/acre/ft} * 1 \text{ ft/12 in.} * 0.01 \text{ in.}$$
$$= 173{,}913.33 \text{ gallons or } 23{,}237 \text{ cubic feet}$$

This is enough water to raise the surface of a 1/2 acre pond by 1 foot (21780 ft^2 in 1/2 acre). In heat islands, the differences in precipitation are more substantial.

Observing Figure 8.7, we see a yearly mean difference of 15 cm between the center of the Detroit/Windsor heat island and the adjoining urban land (95 cm vs. 80 cm). This is 5.9 inches. Using the formula above with 5.9" instead of 0.01", the result obtained is over 102 million gallons of water (for 1 square mile in an average year). This demonstrates the potential for higher runoff within different urban zones.

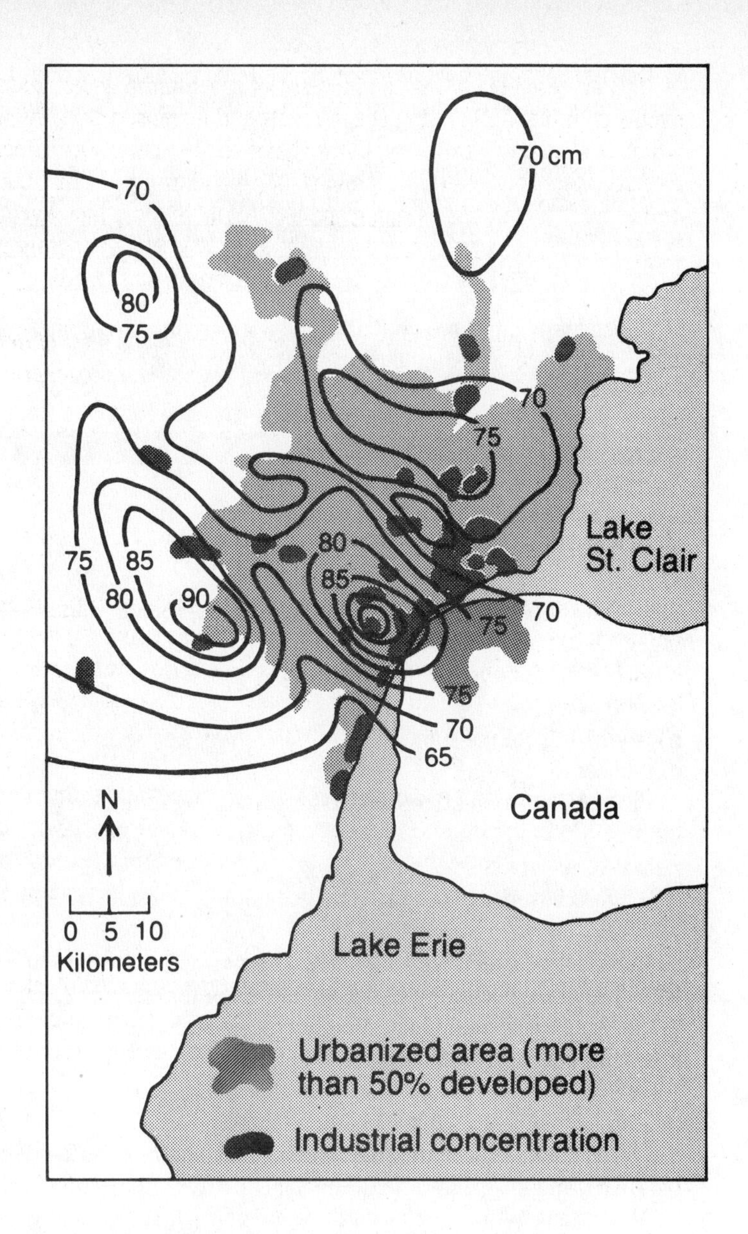

Figure 8.7 Precipitation patterns in the Detroit/Windsor metropolitan area

8.11 The Effects of Urban Climate on Life

Concepts: sensible heat, latent heat

Air pollution and heat have harmful effects on humans. Higher incidence of respiratory ailments and heat-related health problems, including deaths from heat syndrome are found in urban areas. What can be done to ease these problems in the most built up areas of our cities?

sensible heat

Most of the heat released from urban landscapes is sensible heat; this is the heat of dry air and it causes temperatures to rise. If possible, it would be beneficial to convert areas generating sensible heat into latent heat emitters.

latent heat

Latent heat is the heat of moist air, and this type of heat does not cause temperatures to rise. Vaporization of water from plants, soils and water features generates latent heat.

One approach for minimizing the sensible heat generated in urban areas is to plant trees on vacant land in central areas or convert the lots to detention ponds. Since these areas often receive excess rainfall from the heat island effect, this measure has multiple benefits.

Since many cities are not willing to remove the development potential from vacant land, other measures should be tried. Got any ideas?

Applications / Discussion

8.2) The atmosphere produces different sights, sounds, and smells. Explain the following:

a) rainbows
b) "twinkling" of the stars
c) thunder
d) the pungent smell after thunderstorms with lightning

8.3) Observe a fire outdoors. Why does smoke normally rise in the atmosphere?

8.4) Watch the evening weather on television for one week. What types of pressure systems are associated with different types of weather? Why? How might this pattern be affected by a change of seasons?

8.5) Consider this scenario: the sun becomes much cooler (its temperature is reduced by one-half), and the distance between the earth and sun decreases enough to maintain the solar constant. Would life on earth be affected? If so, explain why, and how.

8.6) Why is depletion of the ozone layer such a concern? In what part of the world are the effects of ozone depletion being felt the most?

8.7) Explain how at any given point in time a sub-region on the earth's surface could be out of energy balance.

NO 8.8) Have a class debate on the issue of global warming. One side supports the idea that warming is largely caused by human activity and will continue; the other side believes that blocking factors, such as aerosols and clouds, will offset the temperature gains.

8.9) Pick a sub-process of global climate change and construct a model of that process, e.g., pressure changes, increased methane levels, etc.

8.10) Record the temperatures of the following features in a residential urban area on a sunny day:

a) ambient air temperature
b) concrete driveway, no shade
c) dark roof, no shade
d) short grass, no shade
e) within the foliage of a tall tree (attach the thermometer to a boom)
f) soil in the shade of the same tall tree

What do the temperature ranges tell you about the amount of energy radiated into nearby structures? What design implementations of homes can take advantage of these temperature differences?

8.11) Within a one square mile area in most cities, there are typically many different land uses--commercial, residential, industrial, etc. Using aerial photographs and a land use map, identify the potential for minimizing the sensible heat generated within a selected one mile square area. Look at vegetation patterns, street width, vacant land, depression zones where water might be stored, and other possibilities.

Recommended Reading

EPA Journal (quarterly)

Stren, R., R. White, and J. Whitney, (eds.). (1992). *Sustainable Cities: Urbanization and the Environment in International Perspective*. Westview Press, Boulder, CO.

Sources of Additional Information

INTERNET: National Weather Service Regional Climate Centers

Midwest: kunkel@mccc.sws.uiuc.edu
Northeast: nrcc@cornell.edu
Southeast: sercc@cirrus.dnr.state.sc.us
Southern: krobbins@maestro.srcc.lsu.edu

National Oceanic and Atmospheric Administration, National Climatic Data Center, 151 Patton Ave., Asheville, NC 28801, (704) 271-4682

National Weather Service, National Meteorological Center, Room 805, World Weather Building, Washington, D.C. 20223, (301) 763-4670

References

Changnon, S.A. (1992). "Inadvertent Weather Modification in Urban Areas: Lessons for Global Climate Change," *Bulletin of the American Meteorological Society* (73), 619-627.

Mather, J.R. (1974). *Climatology: Fundamentals and Applications.* McGraw-Hill, NY.

Miller, A., and J.C. Thompson. (1975). *Elements of Meteorology* (2nd ed.). Charles E. Merrill, Columbus, OH.

9

AIR POLLUTION: PATTERNS, TRENDS, AND IMPACTS

Review

9.1 Summary

From a pollution perspective, humans have been particularly cruel to their atmosphere. Perhaps this has resulted from the ability of the atmosphere to quickly transport contamination away from its source. If we were to eavesdrop on the conversations taking place in factories, power plants, and automobiles, we might hear people saying: "Hey, I don't smell anything."

Too often, weather and physical conditions give us a jolt of reality. In urban areas, our health and property suffer as temperature inversions and local topography interact to form dust domes and photochemical smog. "Out in the country," aerial photography and remote sensing allow us to view the broad extent of damage to soil, water, and vegetation from air pollution, such as acid precipitation.

Our response to these problems has been slow, and in many cases, misdirected. Automobile use—a major source of air pollution-- continues to increase in LDCs and MDCs. Alternatives to automobiles have not been thoroughly considered in most countries, and although most people spend more time indoors, the focus of most air pollution evaluations has been almost entirely directed to outdoor air pollution. When we do respond, the focus has been an "end of the pipe" approach, rather than on reducing air pollution at its sources.

9.2 Types of Air Pollution

Concepts: natural pollution, human pollution, primary pollutants, secondary pollutants, acute effects, chronic effects, observation

natural pollution

Pollution of the atmosphere occurs from natural and human sources. Natural sources of air pollution include particulates and carbon dioxide ejected from volcanoes, dust from soil erosion, pollen and spores, materials released from forest fires, and sulfur produced by microscopic ocean plankton. Human activities produce several air pollutants related primarily to fossil fuel combustion, aerosol use, and chemicals used in industry and agriculture.

human pollution

primary pollutants

Human pollutants are characterized as primary or secondary. Primary pollutants enter the atmosphere directly; e.g., from stationary point sources (smokestacks), or from

TABLE 9-1 Human Sources of Major Atmospheric Pollutants

Type/Pollutant	Source
Primary	
Sulfur dioxide	fossil fuel combustion, coal fired power plants, ore smelters
Nitrogen dioxides	fossil fuel combustion, automobiles, power plants
Carbon monoxide	fossil fuel combustion, automobiles, tobacco smoke
Particulates	combustion of vegetation and fossil fuels, lead from automobiles, construction, agriculture
Hydrocarbons	fossil fuel combustion, home heating
Synthetic compounds	aerosols, manufactured organic compounds
Radioactive substances	nuclear operations, uranium decay in rock strata
Heat	fossil fuel combustion
Secondary	
Oxidants	automobiles, interaction of hydrocarbons, nitrogen dioxides, and solar energy, photochemical smog, ozone
Acid Deposition	power plants, sulfur dioxide and nitrogen oxides combine with water in cloud droplets

secondary pollutants

mobile sources (automobiles). Secondary pollutants result from the interaction of primary pollutants aided by an external energy source, such as the sun; photochemical smog and acid precipitation are examples. Table 9-1 summarizes the types of pollutants humans produce by source.

acute effects

All the pollutants above are designated as major because of their significant impact on the environment and human health. They are also widely produced by human activity. The impacts of air pollution on human health are of two broad types: acute and chronic. Acute impacts often accompany stagnant air over urban areas which fails to mix vertically and horizontally. The presence of a major pollutant during these atmospheric conditions can have sudden, or acute health effects on humans, including death. In 1952 in London, England four thousand people died when a stagnant sulfur-laden fog covered the city for several days.

chronic effects

Some urban areas have persistently high levels of pollutants which are not immediately life-threatening, but produce lingering, or chronic effects. Aggravation of asthma and emphysema, and eye irritations are common chronic ailments associated with air pollution.

In general, the ability to observe the effects of pollution has provided an additional spark for public protest and environmental regulation. A major oil spill off the California coast in the late 1960's led to many of the subsequent environmental regulations enacted in the U.S. Air pollution is most likely to generate public concern when the health or environmental effects are visible, but its short durations within the major concentrations of population may be acting as a brake on effective action.

observation

9.3 Sources of Air Pollution in the MDCs and LDCs

Concept: pollution sources

A common factor in many air pollution problems is the automobile. In the MDCs, the high density of automobiles per person leads to high discharges of hydrocarbons and nitrogen oxides (NOx). Hydrocarbons and NOx contribute to photochemical smog, and NOx contributes to acid rain. Within the LDCs, the automobile problem has a slightly different twist. A growing number of the most populated cities are now in the LDCs, and in these cities the amount of pollutants being released remains high because the large populations are compensating for the lower rates of automobiles per person.

pollution sources

As LDCs increase their economic development, car ownership per person is likely to rise. Table 9-2 shows this is happening in two countries of particular interest—China and India. Since almost two-fifths of the world's population is contained within these two countries, worldwide air pollution control and air quality improvement will continue to be difficult.

TABLE 9-2 Population Per Automobile, Selected Countries (1987 and 1991)

Country	Persons per automobile	
	1987	1991
United States	1.8	1.7
France	2.6	2.4
Japan	4.1	3.3
Malaysia	12.4	9.0
Brazil	15.5	12.0
Thailand	94.0	69.0
Egypt	123.6	109.0
Nigeria	141.0	113.0
India	544.4	356.0
China	1,074.0	652.0

Source: Motor Vehicle Manufacturers Association, Facts and Figures. 1993, Detroit, MI.

9.4 Processes, Patterns and Scales of Air Pollution

Concepts: environmental hazards, thermal inversions, proximal scale, distal scale, cross-media pollution

environmental hazards

Air pollution is an environmental hazard; that is, people are at risk from the interaction of human activity and environmental processes. The air pollution risk occurs when humans respire or their eyes are exposed to pollutants in the atmosphere. Environmental processes, including pressure and wind patterns and local topography can prolong the exposure to contaminants in a local region, or transport them long distances.

Dispersal of pollutants depends on horizontal and vertical air motion. Horizontal dispersion is accomplished by wind in the troposphere. Vertical dispersion occurs by convection resulting from the uneven heating of the lower atmosphere; air moves vertically along the pressure gradient.

thermal inversions

When air does not receive adequate horizontal and vertical mixing, thermal inversions often result. Normally, due to the environmental lapse rate, cold air lies over warmer air. When warm air manages to lie over cold air, a stable atmosphere develops and vertical mixing is inhibited. The resulting thermal inversion may be prolonged by local land and water conditions.

Thermal inversions have a variety of causes, and three major types have been identified: subsidence, radiation and advection. A subsidence inversion is caused by a slowly sinking air mass of high pressure. The upper air layers are compressed, and their temperature rises. Frequently associated with the high pressure air mass are plenty of sun and low winds. The sun produces photochemical smog, and the low winds prevent horizontal dispersal. Areas east of the Rocky Mountains in the U.S. are prone to these types of inversions.

In a radiation inversion, terrestrial reradiation on cloudless nights cools the air near the ground more, so the air higher up is warmer. The dense cool air sinks to lower areas, especially down hillslopes. Horizontal dispersion is limited by valley walls, and vertical dispersion by the temperature inversion. These events are generally of short duration, and often affect the cities of Denver, Albuquerque, and Mexico City.

Advection inversions are created when local winds pass over a cool ocean current and inland mountains prevent dispersion. Los Angeles experiences many of these events; here, smog is the rule, not the exception.

proximal scale/ distal scale

One major difference between photochemical smog and acid precipitation is their geographical range. Cities with large volumes of cars and people are the ones bearing the primary burden of their own photochemical smog—the effects occur at the proximal scale. Conversely, acid precipitation has major impacts at the distal scale, often hundreds of miles downwind "out in the country." This feature of acid precipitation makes it tougher to regulate; Canada and the United States have disputed the issue of acid precipitation for many years.

Besides the damage to vegetation, another nasty side effect of acid rain is the damage to surface and ground water supplies. Like all precipitation, acid precipitation follows the pathways of the hydrologic cycle. After reaching the surface, acid precipitation becomes runoff or infiltrates into the soil, where it eventually percolates to groundwater. <u>What are the side effects of acid rain as it follows these two pathways?</u>

Figure 9.1 shows the two major pathways and side effects of acid precipitation. Many of the side effects, particularly in soil, are not clearly understood. There are numerous processes creating the hazard potential, and the hazard is not limited to the atmosphere.

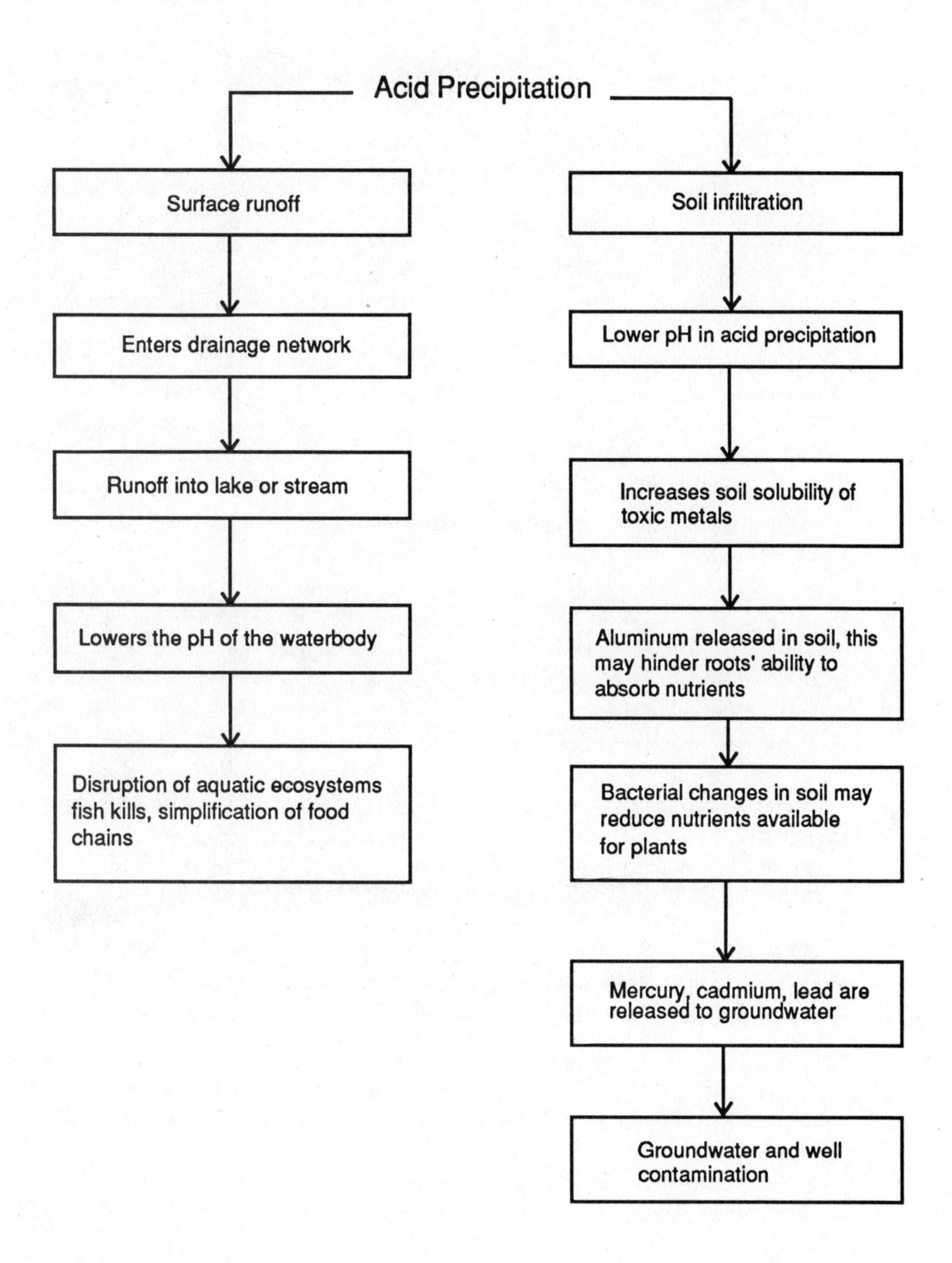

Figure 9.1 Pathways of acid precipitation

cross-media pollution

In the diagram, the multiple pathways of acid precipitation demonstrate an example of cross-media pollution, where more than one medium, such as air, land, or water is impacted. Why would the management of cross-media pollution require a comprehensive, interdisciplinary approach?

9.5 Patterns of Pollution in the Urban Region

Concepts: measurement, total exposure assessment, new building sickness

The severity of the air pollution hazard depends upon a combination of spatial and temporal factors. From a spatial perspective, air pollution usually increases with land use density, particularly with increased concentrations of industrial and transportation-related activities. From a temporal perspective, these pollution generating activities are at their highest levels during the daytime, when the highest concentrations of people are present.

As noted earlier, the air pollution hazard has visible effects. Perhaps this is one reason why most of the environmental regulations enacted by nations have focused on outdoor air quality. Is this a comprehensive approach—is our measurement of the air pollution hazard accurate? The answer is no, if you consider people spend 90 percent of their time indoors in some countries, including the United States.

measurement

How is air pollution measured?

Most air pollution measurements are taken on the roofs of public buildings. Other locations may be used based on their convenience, security, geographic spread, and general match with population distribution. This technique assumes that outdoor air pollution levels are the best indicators of our exposure to contaminants. This is true for some contaminants, such as sulfur dioxide, but significant exposures from indoor sources also occur: such as from tobacco smoke (Vig. 9.1).

Vignette 9.1 Measuring Exposure to Air Pollution

total exposure assessment

Increased recognition of indoor air pollution has led to the concept of total exposure assessment (TEA). This procedure looks at all routes of human exposure to airborne pollutants.

Using TEA, human exposure is measured one of two ways: 1) each pollutant concentration is measured separately in each microenvironment, then the sum is computed by weighting the amount of person-hours spent in each; or, 2) the breathing zone of individuals is sampled several times daily during normal activities.

Small changes in indoor conditions affect total exposures more than large changes in outdoor concentrations. This effect results from the proximity of people to the contaminant. For example, exposure to particulates from the passive smoke of cigarettes is often 50 times higher than total exposure to particulates from power plants. This occurs even though each year power plants produce 500,000 tons of particulates, compared with tobacco smoke particulate emissions of 20,000 tons/year.

Indoor exposure to air pollutants often assumes more subtle forms than tobacco smoke. New building materials are major sources of pollutants. Soon after application, "wet" products such as paints, adhesives, caulks, cleaners, waxes, and polishes emit large fractions of their mass into the building air. After these products dry, they continue to emit very slowly for a very long time.

Follow up

In the past 40 years, the replacement of traditional building materials with "high-tech" products has increased the amount of indoor air pollution. For instance, composite wood products have replaced solid wood materials, bringing binders, adhesives, and other chemical additives indoors.

new building sickness

Increasing numbers of indoor workers have complained of eye and throat irritations, headaches, and lethargy. These are symptoms of "new building sickness," a disorder possibly arising from breathing polluted indoor air. This has prompted regulations from many private businesses and governments to restrict indoor tobacco smoking.

Would you recommend other regulations for improving indoor air quality? Why or why not?

9.6 Air Pollution Impacts on People and Environment

Concepts: environmental hazards, hazard profile

environmental hazards

Several airborne pollutants are damaging to human health. For instance, the rise of particulate concentrations is closely related to increases in hospital admissions for asthma and emphysema; ozone and sulfuric acid (from sulfur dioxide combining with water vapor) damage lung tissue. The risks to health generally increase with age.

Environmental damage from air pollution is also widespread from acid precipitation and deposition of nutrients and residual chemicals. Forests in North America, Japan, and Europe have been damaged by acid precipitation. The Great Lakes in the United States and Canada have experienced atmospheric nutrient loading (phosphorus and nitrogen), and pesticides banned in the U.S. over 20 years ago, such as DDT are still finding their way into the food chains.

hazard profile

The severity, or social magnitude of hazards has always been difficult to determine. For instance, which hazard has more social magnitude: widespread property damage from a hurricane—with no deaths or injuries, or the death of a child struck by lightning during a thunderstorm? Given this difficulty, geographers have developed a classification of environmental hazards called the hazard profile. This method is based on spatial and temporal characteristics, and while it does not indicate which hazard is "worse," it allows planners and policy makers to evaluate different hazards more objectively.

A profile of environmental hazards independent of their human impact is shown in Figure 9.2.

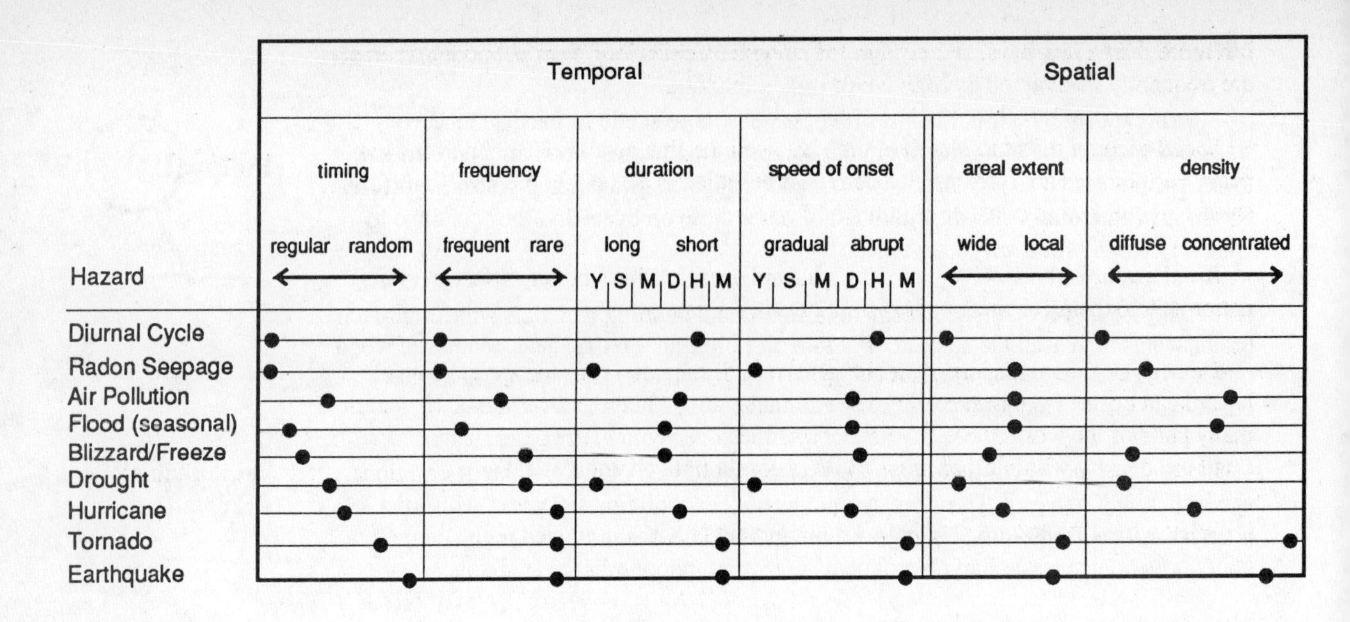

Y = years
S = seasons
M = months
D = days
H = hours
M = minutes

Figure 9.2 Environmental hazard profile

Like all environmental hazards, air pollution has spatial and temporal characteristics, such as frequency of occurrence and areal extent. In the figure, air pollution and other environmental hazards are shown within the context of six temporal and spatial characteristics. There are hazards of regular occurrence and high frequency (the diurnal cycle), as well as low frequency events with poor predictability (earthquakes). Let's examine air pollution within this framework, and assume we are only considering proximal urban air pollution (photochemical smog).

Timing is the sequence of events associated with a hazard. Many events follow a random time distribution (e.g., earthquakes), while others such as floods are seasonal. The timing of air pollution is influenced by meteorological conditions, which are random, and local population densities and physiography, which are known. The timing denoted by the small circle is a compromise between these random and predictable factors.

Frequency of hazards is their recurrence interval based on a long-term average. In a given area, if long-term observations of hurricanes indicate they occur once in every 10 years on average, then in any year there is a 10 percent chance of a hurricane occurring. For air pollution, the average value would depend on the local population densities and climatic patterns, so a compromise value was selected again.

Duration refers to the length of time a hazard persists. Frosts may last for hours; floods can last weeks; droughts often persist for years. Severe air pollution events rarely last more than a few days, and even in the smoggiest cities long term pollution patterns are frequently interrupted by high winds or precipitation.

Speed of onset refers to length of time between the first appearance of an event and its peak. Earthquakes are fast-onset hazards. Air pollution is also relatively fast—damaging smog can build up in a matter of hours or days.

Areal extent is the space covered by a hazard event. A drought may affect an entire nation; photochemical smog covers smaller urban areas.

Density, or spatial dispersion, is the pattern of distribution over the space in which a hazard can occur. For instance, there is a "tornado alley" in the midwestern U.S., and in many parts of the world there are distinct seismic zones with high earthquake activity. Land use density within urban areas tends to concentrate photochemical smog.

<u>What would the hazard profile look like for acid precipitation?</u>

9.7 Air Pollution Control: How Well Are We Doing?

Concept: source control

The success of global air pollution control efforts has largely depended on the quantity of fossil fuel released to the atmosphere. While reductions in sulfur dioxide emissions have been achieved by power plants switching to low sulfur coal, increases in vehicle miles traveled are keeping ozone and carbon monoxide levels high in many MDCs (Figure 9.3). Air quality is also deteriorating in many LDCs as automobile use increases.

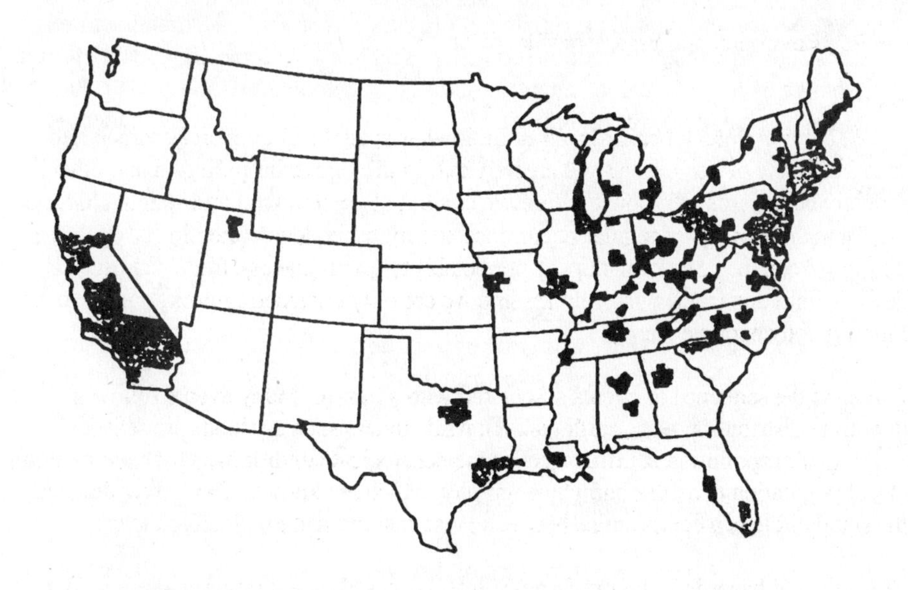

Figure 9.3 Areas violating ozone standards (1991)

The shading represents areas exceeding U.S. national ambient air standards (the allowable amount present in the outdoor air at a given time) for ozone. The shaded areas correspond to major population centers; over 130 million people are exposed to excess ozone.

source control

Regulation of pollution by governments takes two general forms: source control and "end of the pipe" controls. Source control prevents the production of potential pollutants; end of the pipe treatment tries to reduce pollution after it has been generated by a source—but before it reaches the environment. An example will clarify this distinction.

Incineration of garbage often releases harmful by-products into the atmosphere. High levels of dioxin (an extremely toxic chlorinated hydrocarbon) and mercury have been associated with large incineration facilities in several countries. Source reduction applied to incineration would remove items from the waste stream which might produce dioxin and mercury before they are burned—there is no release of toxins. End of the pipe controls use scrubbers or filters to reduce or remove the contaminants after incineration.

Another problem with end of the pipe controls are the test procedures used to measure the pollution released into the environment. Vignette 9.2 discusses this problem in relation to automobile emissions.

Vignette 9.2 "Why Cars Aren't as Clean as We Think," by Marc Ross

Regulations have supposedly made cars a lot cleaner in the last 25 years. Emissions of carbon monoxide and hydrocarbons are nominally a mere 4 percent of their late-1960s levels. But emissions are actually 5 to 10 times higher than currently "permitted" levels. Why is there such a disparity—and what does it imply for policymakers?

One well-known problem is the number of old cars still on the road, manufactured under less stringent standards, but this problem is exaggerated. Some 10 percent of more recently built cars have developed engine or exhaust system malfunctions; this is believed to be the largest source of excess emissions. To address this, carmakers could identify types of malfunction and redesign new vehicles to reduce the likelihood that they occur. But there is another significant and little recognized source of the discrepancy: the official method of measuring emissions is grossly misleading.

In the Federal Test Procedure (FTP), stationary vehicles are driven on rollers under laboratory conditions, following a prescribed second-by-second sequence of speeds intended to simulate a typical urban driving cycle. The exhaust is collected and chemically analyzed. But the FTP driving cycle virtually ignores the episodes of driving that contribute disproportionately to overall emissions: the bursts of high power needed to climb hills and accelerate at high speeds.

The heart of the emissions-control system in today's vehicles is a catalytic converter in the exhaust line. For the catalyst to be effective, the fuel-air mixture delivered to the engine must have the chemically correct ratio; that way, the catalyst can work on the exhaust that has balanced residues of partially burned fuel and oxygen. To maintain this chemical balance, an oxygen sensor is installed in the exhaust line, and its signal is used to adjust the mix of fuel and air entering the engine.

But when high power is called for, today's vehicles override this sensor and inject excess fuel into the cylinders—typically 20-30 percent more than what is needed to chemically balance the oxygen present. This excess fuel raises by orders of magnitude the emission rate of carbon monoxide and hydrocarbons; each second of driving with the pedal down corresponds roughly to 30 minutes of carbon monoxide emissions, and 1 minute of hydrocarbon emissions in the moderate driving represented by the FTP test. And such high-power episodes are not rare. Drivers of small cars, in particular, often find themselves keeping up with expressway traffic, or climbing hills, by driving for long stretches at wide-open throttle.

The design that relies on excess fuel for high-power operation is a vestige of the days of carburetor engines, which required the rich mixture to avoid engine hesitation. With simple changes in engine design, cars could probably achieve high power much more cleanly. European manufacturers have taken a step in this direction; in some new Volvos, for example, high-power emissions are one-third to one-half lower than in most other cars. Modifying the software that controls the air-fuel mix enables these cars to enrich the fuel much less than U.S. cars do, or for a briefer period.

With rare exceptions, U.S. automakers appear not to have seriously studied such approaches, much less implemented them. Of course, government regulators could encourage such design changes by amending the FTP so that it accounted for high-power driving. Carmakers would then be free to devise their own engineering solutions, as long as the met emission standards under truer tests. Unfortunately, some regulators in the United States seem not to understand the problem. California, New York, and a few other states are simply lowering the permissible emissions for cars on the standard FTP cycle. The federal government plans to do the same by 2003. Although well-intentioned, this is bad policymaking, because most of the automotive emissions are not related to those measured in the FTP.

The irrelevance of today's FTP reflects a failure—by government, university, and auto industry labs—to adequately study the fundamentals of automotive emissions. Considering the importance of vehicle emissions and the billions of dollars in control costs borne by vehicle manufacturers and buyers, this neglect is mind-boggling.

Sorting out the physical sources of excess emissions, such as high-power operation, will also allow us to more rationally focus transportation R&D. Ultimately, we will develop alternatives to the internal combustion engine, such as vehicles powered by batteries, fuel cells, or some hybrid of a gasoline engine and electrical energy storage. Large subsidies would be needed to accelerate the introduction of any of the new technologies, because they will, at least initially, cost more and perform worse that what we already have.

But we might vastly and quickly improve what we already have. Addressing the problem of high-power emissions and malfunctioning vehicles could reduce emissions of conventional automobiles by 80 percent or more. That represents a target too fat to ignore.

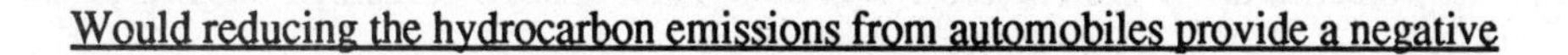

The auto emission testing currently used in the United States demonstrates invalid measurement techniques. In order to be valid, a measurement technique must accurately measure the process generating the quantity under investigation.

<u>Would reducing the hydrocarbon emissions from automobiles provide a negative</u>

incentive for developing alternative power sources, such as electric batteries?

One form of source control applied to automobile emissions involves the reduction of vehicle miles traveled. Using bicycles for short trips in urban areas could make a significant impact. More than half of all commuting trips in the U.S. and nearly three-fourths in the United Kingdom are eight kilometers (5 miles) or less. Bicycling is the most energy efficient form of transportation (Table 9-3), and with the exception of CO_2 from respiration, no hydrocarbons are emitted.

TABLE 9-3 Comparison of Energy Intensity for Selected Transport Modes, United States, 1984

Mode	Calories per passenger mile
Automobile	1,860
Transit Bus	920
Transit rail	885
Walking	100
Bicycling	35

Source: Mary C. Holcomb et al. (1987). Transportation Energy Data Book: Edition 9, Oak Ridge, TN.

Which method of pollution control is more effective—source control or end of the pipe?

Source control is more cost-effective and less damaging to the environment. Generally, less capital investment is needed for pollution control equipment, and since the pollutant is never produced, there is no release to the environment. A major difficulty associated with source control is the need to change attitudes in industry and among the population about the products which are manufactured and used.

In urban areas, reducing inputs of transportation-related energy provides the most pollution control benefits. There are good opportunities in all countries, especially within the tropical and sub-tropical LDCs to incorporate these energy saving measures into future urban plans. We can do better.

Applications / Discussion

9.2) Perform an air pollution inventory of your home. Using the outline provided by Table 9-1, record the potential sources of air pollutants. Gas furnaces and gas burning stoves emit nitrogen dioxide. Are there smokers in your household? Did you look at all those containers under the sink, or in the linen closet? And the garage...

9.3) Many LDCs often have more buses than cars. Buses give off more pollution than cars, but the rate of car ownership is rising rapidly in many nations. Given an "either/or" choice, should policy-makers in LDCs try to encourage bus usage, or work to reduce the amount of new cars being purchased? What information would help you make a recommendation?

9.4) Which type of temperature inversion is your community most likely to experience? Why?

9.5) Using newspapers which print precipitation and pollution statistics (amount of rain/snow per day, pollen counts, particle counts, etc.), construct a graph of particulates on the y axis vs. precipitation amounts on the x axis. What relationship is there between airborne particulates and precipitation amounts?

9.6) Using Figure 9.2, construct a hazard profile for indoor air pollution.

9.7) What potential reductions of air pollution are there in your community from bicycles or other substitutes for automobile travel? Is there a large employer centrally located? Are there access lots available for shuttle parking?

Recommended Reading

Garreau, Joel. (1991). *Edge City: Life on the New Frontier*. Doubleday, NY.

Nadis, S., and J.J. MacKenzie. (1993). *Car Trouble*. Beacon Press, Boston, MA.

Popkin, R. (1989). "Source Reduction: Its Meaning and Its Potential," *EPA Journal* (15), 27-29.

Sources of Additional Information

Air Resources Information Clearinghouse, 46 Prince St. Rochester, NY 14607, (716) 271-3550

Center for Clean Air Policy, 444 N. Capitol St., Suite 602, Washington, D.C. 20001, (202) 624-7709.

INTERNET: Several Air Pollution Gophers on the North American Menu

References

American Automobile Manufacturers Association. (1993). *Facts and Figures*. Detroit, MI.

Burton, I., R.W. Kates, and G.F. White. (1978). *The Environment as Hazard*. Oxford University Press, NY.

Elsom, D.M. (1992). *Atmospheric Pollution: A Global Problem*, (2nd ed.). Blackwell Scientific, Cambridge, MA.

Holcomb, Mary C. (1987). *Transportation Energy Data Book: Edition 9*. Oak Ridge, TN.

Koutrakis, P. (1990). "Acid Air and Health," *Environmental Science and Technology*, (24). 946-956.

Smith. K.R. (1993). "Taking the True Measure of Air Pollution," *EPA Journal* (19), 6-8.

10

THE HYDROLOGIC ENVIRONMENT AND LAND USE

Review

10.1 Summary

The management of water provides one of the greatest challenges on earth. How can we redistribute its vast quantities and insure an ongoing stable supply?

Water is a renewable resource—this is in our favor. Solar energy spawns the hydrologic cycle of precipitation, runoff, storage, and evaporation which constantly replenishes our fresh water supply. But access to this water is constantly being affected by poor management and pollution.

Each year, thousands of floods occur which produce excess water. Floods are nature's way of saying: "Here is excess water—take it, and use it wisely." But rather than viewing this water as an asset, society has sought to control and limit these events. These attempts have often led to more severe floods, with continuing death and destruction, while leaving little to show in terms of water salvaged for future use.

Wasteful practices of water management have evolved through our historical experiences, in which we have always tried to control water with technology, rather than treat it as an integrated part of the ecosystem. As society marveled at its technological triumphs over raging streams, these same streams became dumping grounds for human waste. Successful water management will arrive when humans manage their thirst for water as an integrated whole with the processes affecting soil, the atmosphere, and other living organisms.

10.2 The Hydrologic Cycle

Concepts: hydrologic cycle, fresh water supply

hydrologic cycle

The hydrologic cycle is the system linking earth's water resources—the oceans, lakes, streams, groundwater, soil moisture, and atmospheric vapor. In this cycle, pathways for water in its gas and liquid phases traverse all spheres of the geosphere, and consist of processes initiated by external energy sources. For instance, two primary pathways within the hydrologic cycle are between the surface components of the hydrosphere (oceans, lakes, streams), and the atmosphere. Heat energy from the sun causes evaporation and the condensation of water vapor. Gravitational force helps to transport the water to land via precipitation, and also allows rivers to replenish the oceans (Figure 10.1).

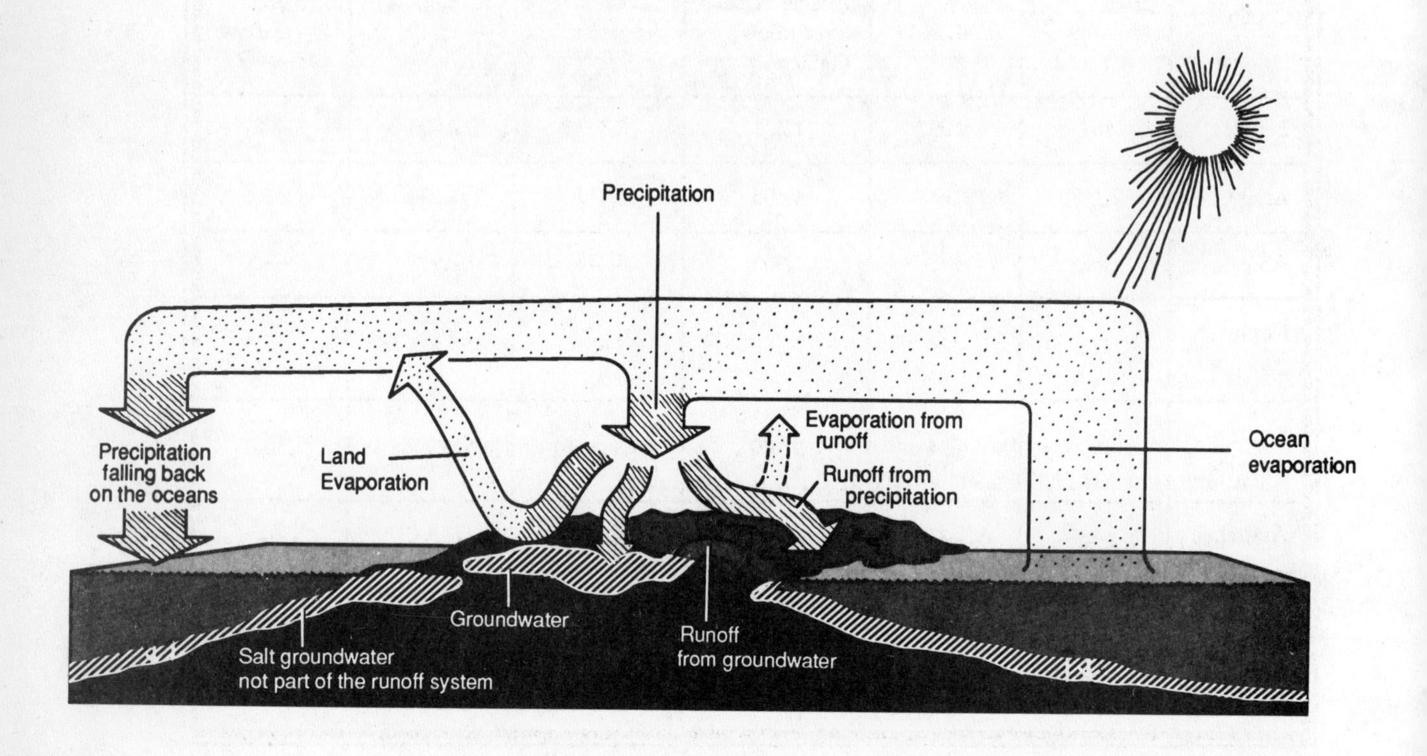

Figure 10.1 The hydrologic cycle

fresh water supply

The hydrologic cycle works like a large desalinization plant. When salt water from the oceans is evaporated, the precipitation produced falls as salt-free fresh water on the oceans and continental land masses. Fresh water is stored in glaciers, rivers, lakes, and groundwater; with surface water being the most easily accessible.

To get an idea of how well the people of the world are supplied with water, we can examine the per capita fresh water resources. Streamflow is a good indicator of supply since it is the most accessible and easiest to estimate in terms of volume. Table 10-1 compares the 1970 and 1990 per capita stream water resources for the inhabited continents.

The data demonstrate two important concepts: 1) the easily accessible fresh water on earth is distributed unevenly; and, 2) while there is still an adequate amount of water per person on all continents, increasing population has made significant reductions in the

TABLE 10-1 Per Capita Continental Water Resources: 1970 and 1990

Continent	Population 1970 (millions)	Population 1990 (millions)	Average Annual Stable Streamflow (km3/yr)	1970 Stable Streamflow per capita (m3/yr)*	1990 Stable Streamflow per capita (m3/yr)	% Change in Stable Streamflow per capita
Europe	467	509	1,325	2,837	2,603	-8.2
Asia**	2,338	3,399	4,005	1,713	1,178	-31.2
Africa	363	643	1,905	5,248	2,963	-43.5
North America	318	424	2,380	7,484	5,613	-25.0
South America	191	294	3,900	20,419	13,265	-35.0
Australia	15.4	20.5	495	32,143	24,146	-24.9

* one cubic meter equals 264.2 gallons ** includes former USSR

Sources: Lvovich, M. (1973). The World's Water. Mir Publishers, Moscow; United Nations Population Yearbook, 1994.

amount of fresh water readily available in the LDCs, particularly in Asia and Africa.

Despite the seemingly large amount of fresh water available, we cannot become complacent. Shortages often occur in areas with ample precipitation. In these and other areas, pollution has already limited the fresh water supplies, and this, along with the uneven geographic distribution of fresh water supplies and their mismanagement (discussed later), have limited our access to this resource.

10.3 Water Vapor, Condensation, and Precipitation

Concepts: precipitation process, precipitation forms

precipitation process

The precipitation process involves the cooling of water vapor and the ensuing growth of water droplets around condensation nuclei. Small water particles are formed when the atmosphere is cooled—this is condensation. These small droplets are then attracted to small particles with an average size of 0.001 mm in the atmosphere called condensation nuclei. The condensation nuclei consist of suspended particles from combustion, volcanic activity, soil, and meteoritic dust, and they provide a site for water droplet growth. When the droplets reach a mass large enough to fall, they become precipitation.

We are not sure how precipitation droplets grow, but the amount and rate of growth are impressive. In a non-precipitating cloud, the average drop diameter is only 0.02 mm. Raindrops at the earth's surface are usually between 1-10 mm, which means that most cloud drops have increased their size by a factor of 100, and their volume by a factor of 1 million. This is shown in table 10-2.

TABLE 10-2 Differences in Precipitation Droplet Volume

For a droplet 0.02 mm in diameter:

radius, r = 0.01 mm   volume = $\Pi r^3 = (3.14 * 0.01)^3 = 0.000031$

For a droplet 2.0 mm in diameter:

radius, r = 1.0 mm   volume = $\Pi r^3 = (3.14 * 1.0)^3 = 30.959144$

Calculating the droplet volume difference:

30.959144 / 0.000031 = 1,000,000

The 2.0 mm diameter droplet has 1 million times the volume of the 0.02 mm diameter droplet

The period between cloud formation and precipitation is often less than one hour. This high rate of cloud droplet growth cannot be explained solely by the condensation of water vapor on existing water particles—the process would take too long. Scientists believe the rapid increase in droplet size is attributable to two possible mechanisms: collision and coalescence of particles, and the growth of ice crystals.

Collision and coalescence of particles is promoted by the differing water particle sizes within clouds. Particles of varying sizes exist because condensation occurs at different rates in clouds. Swirling air in the cloud creates many different paths for the larger droplets, and collisions between the larger and smaller drops occur. The smaller and larger drops coalesce (combine) and form very large drops. These very large drops splinter into several drops, which in turn grow by collision and coalescence—there is a chain reaction of raindrop growth.

Although not completely understood, ice crystals seem to stimulate precipitation by mixing with supercooled liquid drops. In the middle latitudes, the top layers of clouds are often composed of ice crystals, and the lower layers contain supercooled (below 0° C.) liquid drops.

Wait a minute. How can water exist as a liquid below 0° C? In large volumes of water such as lakes and ponds, freezing does occur at or near 0° C. But in very small droplets water may not freeze until it reaches -40° C. In the atmosphere it is common for liquid water drops to exist in clouds at temperatures as low as -20° C. Why does this happen?

For pure water to become ice, the molecules must assume a particular configuration. The exact arrangement is reached through chance, and the probability of the correct arrangement being reached increases with the volume of water involved. This is why very small particles do not freeze until they reach a temperature below the "normal" freezing point of 0° C. From this, it would be more accurate to refer to 0° C. as the melting point of water.

Now let's return to the mixing of ice and supercooled water.

Air currents within the cloud, or different fall velocities of the particles cause ice crystals (from the top layers) and the supercooled drops (from the lower layers) to mix. At the same temperature, ice crystals have less saturation vapor pressure than the liquid water—in effect, there is no more room within the ice for gaseous water, so the ice crystals expand. This growth is believed to be important in the initiation of precipitation, and is called the Bergeron process after the Swedish meteorologist who first suggested it.

precipitation forms

Regardless of the mechanism operating to spur precipitation, when a certain mass is reached, water particles fall to the surface in different shapes, sizes, and states. Table 10-3 summarizes the characteristics of precipitation.

TABLE 10-3 Forms of Precipitation

Type	Characteristic shape; size (diameter); state
Drizzle	droplet; 0.05 mm (0.02 in.); liquid
Rain	droplet; greater than 0.05 mm; liquid
Freezing rain/drizzle	droplet; rain or drizzle which freezes on impact with the ground or objects
Snow pellets	spherical or conical; 2-5 mm (0.01-0.2 in.); opaque white ice grains
Snow grains	flattened shape; <1 mm; white opaque ice grains
Sleet	semi-spherical; <5 mm; transparent or translucent; frozen raindrops or snowflakes
Hail	circular; 5-75 mm (0.2-3.0 in.); layered ice shells with an onion-like cross section

Source: Miller and Thompson. (1975).

10.4 Precipitation, Storms, and Their Impact on Environment

Concepts: types of precipitation, magnitude and frequency

types of precipitation

Four different cooling mechanisms produce four general types of precipitation: convectional, orographic, frontal/cyclonic, and convergent. A useful way to distinguish between each type of precipitation is to list their environmental impacts (Table 10-4).

TABLE 10-4 Environmental Impacts Associated with the Four Types of Precipitation

Precipitation Type	Cause	Environmental Impacts
Convectional	surface heating and thermal formation	lightning, fires, soil erosion
	intense thermals	lightning, fires, wind damage, soil erosion, flash flooding
Orographic	prevailing winds against mountains	high amounts of runoff, severe erosion
	temperature inversions	air pollution, acid fog, vegetation and lake damage
Frontal/Cyclonic	interplay of warm/cold air masses, coriolis effect	warm front--light rain, some soil erosion
		cold front--lightning, hail
		tornadoes--crop/property damage, regional systems can produce large floods
Convergent	warm moist air is drawn into areas of low pressure	monsoons--flooding, severe soil erosion
		hurricanes--floods, property damage

magnitude and frequency

Hurricanes are considered to have the most damaging impacts; this is based on their magnitude (total energy released). On the basis of frequency, thunderstorms are the worst since they produce an estimated 3 billion lightning strikes per year, resulting in more deaths than all other storm types and causing many fires in forested and urban areas.

Why do we rarely take special precautions for thunderstorms?

10.5 Streamflow and Floods

Concepts: baseflow, stormflow, hydrographs

baseflow

There are two components to streamflow: surface runoff from rainfall and snowmelt, and groundwater seepage into the stream channel. The most consistent source of water comes from groundwater, which feeds streams in dry and wet weather. This supply of water is called baseflow.

stormflow

Surface runoff contributes water to streams only during heavy rainfall and snowmelt events. This part of the stream's discharge is called stormflow. Flood magnitudes and the impacts from pollution increase as stormflow increases.

hydrographs

Streamflow is thus a combination of overland flow (stormflow), and baseflow, and is commonly expressed in units of cubic feet per second (also called discharge, and abbreviated as "Q"). In the United States, the U.S. Geological Survey measures streamflow at about 9000 gauging stations. One method used to study streamflow is the hydrograph, which is a plot of streamflow discharge against time (Figure 10.2).

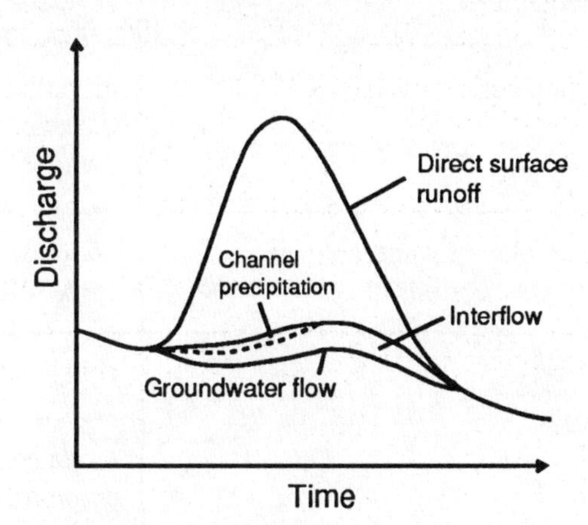

Figure 10.2 The hydrograph

There are four components to a hydrograph: direct surface runoff, interflow, baseflow, and channel precipitation.

Direct surface runoff consists of water flowing overland until it reaches a stream channel. During large storms, this is the most significant hydrograph component.

Interflow is subsurface flow moving at shallow depths and reaching the stream channel in a relatively short period of time.

The baseflow component is water percolating downward until it reaches the groundwater reservoir; over longer periods of time this water reaches the surface stream as groundwater discharge. During the hydrograph time period, baseflow increases are not usually detectable.

Channel precipitation (rain falling on the stream) begins with the onset of rainfall and ends with the storm. Since most streams occupy less than 5 percent of total watershed area, this component contributes only a small fraction of the total discharge during a storm.

Hydrographs are useful tools for examining floods and the responses of different watersheds to storms. The hydrograph of the stream at bankful stage defines the discharge required to produce a flood. In terms of watershed response, peak flows may change with increasing urbanization, and the time it takes to reach the peak flow largely depends on a combination of storm intensity, land cover, soils, slope, and land uses within the watershed. Comparison of hydrographs for the same stream location over time can lend insight into the effects of changing land use within a watershed.

What would be the dominant surface feature of a watershed which exhibited very little rise in the direct surface runoff component of the hydrograph during a major storm event?

10.6 Flood Hazard and Land Use

Concepts: floods, flood control and ecosystem impacts

floods

Floods take three major forms: 1) those which are riverine (river) related—flash, alluvial fan, meandering stream channels, and ice jams; 2) flooding outside the normal floodplain from surface runoff; and, 3) coastal flooding.

In general, riverine floods occur when water overflows the channel and enters the adjacent floodplain (Figure 10.3).

Figure 10.3 Riverine flooding

Flash flooding occurs in narrow, steep valleys, and along urban drainage courses. In the United States, over three-fourths of all Presidentially declared disasters involve flash flooding; and flash floods have caused the most weather-related deaths.

Alluvial fan flooding occurs at the base of mountain ranges. These floods can cover wide areas, and often have floodwaters with high velocities resulting from the slope of the stream as it descends the mountains and nears the valley floor.

Meandering stream channel floods can result from human activities such as mining, overgrazing, forestry, dam building, and urbanization. Surface water patterns change in response to these activities, and this makes the potential areas affected by these floods difficult to predict.

Ice jams cause a rapid rise of water at the point of the obstruction and upstream. When the jam breaks, downstream flooding results. Additional damage is often caused by the force of the ice.

Flooding outside the normal floodplain is most often related to inadequate drainage due to urbanization or frozen ground.

Coastal flooding results from storm surge and fluctuating lake levels. Storm surge is the rise in the water surface from a change in barometric pressure and the piling up of water as a result of wind. Wave action also contributes to storm surge. Lake levels change over long time periods in response to changes in precipitation, runoff, and evaporation. Over shorter time periods, lake levels may be affected by strong winds, barometric pressure changes, and water diversions.

In the United States, despite increased investment in flood control, damage from floods has increased. Flood control efforts exhibit a wicked paradox, as many of the engineering-based flood control efforts have actually increased the magnitudes of certain floods. The 1993 Mississippi River flood cited in *Environmental Geography* is a good example of how navigation and flood control facilities have actually contributed to this phenomenom.

flood control and ecosystem impacts

There are also hidden costs of engineered flood control methods, particularly the long-term impacts they have on ecosystems. The most damaging is channelization—a group of engineering practices used to control flooding and facilitate navigation; activities include channel straightening, widening, and deepening.

Channelization actions affect the physical and biological environments within the stream and on the adjacent banks. Channel straightening leads to an increase in flow as the straightened channel gouges out a wider and deeper path. This destroys habitats, since many aquatic species have specific requirements of water velocity. When bankside vegetation is removed during channelization, there is a loss of cover for many aquatic organisms and an increase in water temperature. These events generate secondary feedback effects within ecosystems such as fish kills and over-predation.

10.7 Groundwater

Concepts: subsurface water, aquifers, sustainability, conjunctive use

Figure 10.4 shows the zones of water which exist beneath the earth's surface.

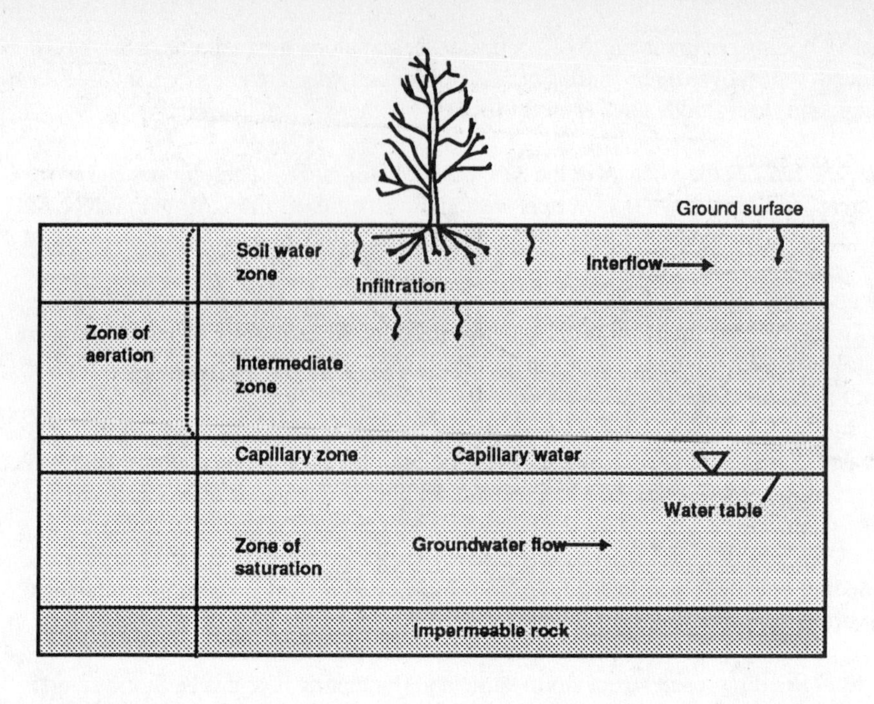

Figure 10.4 Subsurface water zones

The amount of water generally increases as you go deeper into the ground. In the zone of aeration, the pores between rock and soil are filled with water and air. In the zone of saturation, the rock and soil pores are completely filled with water. At a greater depth, the water table is the surface of the zone of saturation. It is under atmospheric pressure, and changes elevation depending on the net recharge of the groundwater system. When discharge exceeds recharge, the water table level falls; the level rises when recharge is above discharge.

aquifers

Aquifers are geological formations that will yield enough water to be economically used and developed. There are two main types of aquifers: unconfined and confined (Figure 10.5). In an unconfined aquifer the top of the zone of saturation (the water table) is not restricted by an impermeable layer of rock and is at the same pressure as the atmosphere (approximately 1013 mb). If a well is drilled into an unconfined aquifer, the water will rise only to the level of the water table. This is why many wells need pumps to bring the water to the surface.

Confined aquifers, or artesian aquifers exist between impervious layers of rock. This causes the water within them to be at a higher pressure than the atmosphere. When a well is drilled into a confined aquifer, the water in it will rise above the surface.

sustainability

In many parts of the world, the water tables of aquifers are being lowered by overuse. Concern over the depletion of groundwater supplies has led to the concept of safe aquifer yield. This is really a type of sustainable use, since the objective is to preserve a long term balance by maintaining pumping rates which do not exceed the recharge rates of the system.

conjunctive use

In order to use water resources efficiently, the characteristics of surface and

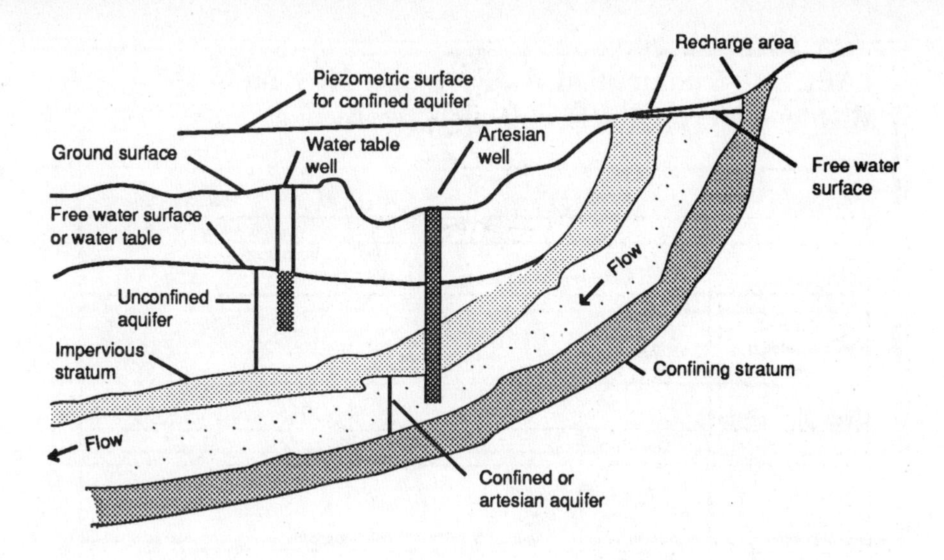

Figure 10.5 Types of aquifers

subsurface water should be managed together. Although surface water and groundwater systems are physically inseparable, these systems are often managed independently. In the United States, few, if any states have laws which protect or specify the allocations of surface water and subsurface water jointly. One method for achieving this is through conjunctive use.

Conjunctive use has been applied cyclically in semi-arid areas where water conservation is crucial. One such conjunctive use system is used in the Los Angeles Coastal Plain in California. During the seasons when precipitation is most available, surface waters are collected and used to recharge aquifers. In the dry seasons, the stored groundwater is used, with the knowledge the water table will be replenished in the next wet season.

<u>What risks are involved with this type of conjunctive use in California</u>?

10.8 Lakes, Ponds, and Wetlands

Concepts: water budget, eutrophication, water quantity/quality relationships

In areas where the water table intersects the ground surface, groundwater is exposed. In valleys, the water is carried away by streams. In low spots where an outlet to a stream is limited or nonexistent, seepage water accumulates, and natural impoundments form—lakes, ponds, and wetlands.

water budget

Natural impoundments exhibit a balance between inputs and outputs of water. Incoming water in the form of precipitation, runoff, or streamflow is released via outputs to streams, evaporation, transpiration, and ground seepage. One method used to study this balance, and to help identify and assess the impacts of human activity on water resources is the water budget. This is similar in practice to the budgets used in financial accounting. In a water budget, the two sides of the "balance sheet" consist of all the inputs (inflows) and outputs (outflows) to an impoundment. Table 10-5 shows an abbreviated water balance for the lakes Michigan and Huron within the Great Lakes of the north central

TABLE 10-5 Simplified Water Budget for Lakes Michigan-Huron (in cfs, cubic feet per second)

Inflows		Outflows	
Precipitation	109,000	River discharges	187,000
Surface runoff	90,000	Evaporation	87,000
River discharges	78,000	Chicago diversions	3,000
Totals	277,000		277,000

U.S.

A more comprehensive water budget would consider other inflows and outflows, such as groundwater seepage into and out of the lake and evapotranspiration. Because this technique requires the identification of the sources and sinks of water at a given scale, it is often used as the first step in the environmental analysis of the human impacts on specific water resources.

eutrophication

For instance, on a hypothetical impoundment, several water budgets performed over many years show increased diversions by cities, and increased discharges from their sewer treatment facilities. Comparison with other data links the increased wastewater discharges to algae blooms and increased overgrowth of aquatic vegetation. These events are symptoms of eutrophication—excess nutrient loading.

During algae blooms, the waterbody becomes saturated with oxygen from the photosynthesis of the organisms. As the algae age and die, they sink and are decomposed by bacteria which use oxygen in the process. This decomposition process can deplete the stored oxygen in the waterbody and create a condition of decreased oxygen called oxygen sag. Below certain levels of oxygen, certain aquatic species may die. Trout, for instance, require at least 5 mg/l of dissolved oxygen to survive. This is why trout streams are characteristically cold (oxygen absorption increases as water temperature drops) and algae-free.

water quantity/ quality relationships

In this example, the relationship between the volumes of sewage discharge and eutrophication demonstrates the linkage between water quantity and water quality. This linkage requires joint management, but is often overlooked. Let's look at one example where the quantitative and qualitative aspects of water management are performed together.

Some basins retaining stormwater in urban areas exhibit the principle of simultaneous management of water quantity and quality. Related to quantity, if properly sited, excess flows are stored in the basins, helping to mitigate floods. Related to quality, during the storage period some pollution-carrying particles in the runoff do not reach the nearest stream or lake because some settling occurs in the basins.

10.9 Human Use of Water

Concepts: water use efficiency, water quantity/quality relationships, scale

Throughout history, people have tried to change the distribution of water resources to their advantage. Irrigation is as old as stable agriculture, and aqueducts were built thousands of years ago to supply fresh water to Roman cities.

As with other human activities, both obtaining water and using it have significant environmental impacts. Because of the uneven distribution of fresh water, significant electrical energy is often used to transport water from distant dams and reservoirs to agricultural or urban areas where water is scarce. In the American southwest, water is transported hundreds of miles to irrigate the agricultural regions in California and supply drinking water for Phoenix and Tucson.

You might ask: is all this effort really worth it? At first glance the answer is yes, because the cities are already there, and the agricultural areas contribute a substantial portion to the national food output.

water use efficiency

If we take a harder look and ask: Are these efforts efficient?, the answer is no. We are wasting huge quantities of water in agriculture (which uses the majority of fresh water), while expending great efforts on domestic uses which are relatively minor from a quantitative perspective.

water quantity/ quality relationships

Conserving water is important because all water use degrades its quality. It therefore makes good sense to focus conservation efforts on those sectors of the economy which use the most water. Runoff from agriculture is a significant contributor to stream pollution worldwide, yet is exempt from any pollution control legislation in the U.S. and most other nations. More efficient irrigation techniques, that is, those which deliver a higher percentage of water directly to the plant, would save water and also reduce pollution from runoff. Large scale dams not only disrupt the ecosystem of the river, their reservoirs often experience high rates of evaporation—another waste of water.

There are other examples of massive water waste. We have already discussed the subsidies of water to large farms in the U.S. west (section 6.10). Conjunctive use (section 10.7) would also help water-poor areas conserve their water supplies.

scale

At a smaller scale, water conservation in private residences can also improve water quality. About 70 percent of domestic fresh water consumption ends up as wastewater. Increasing the efficiency of the two major domestic uses—toilet flushing and lawn watering can have significant impacts on local water quality. Using alternate days for lawn watering, and watering in the early morning can save about 25 percent of the total water used by a typical residence in North America. Retrofitting plumbing fixtures with low-flow devices and using low flush toilets—which use 1.6 gallons of water per flush compared to 6.5 gallons in regular toilets--can save another 15 percent. A forty percent reduction in water use translates into 28 percent less wastewater from domestic sources ($40 * 0.70 = 28$).

Saving water provides many examples of the relationship between water quantity and water quality. If the changes involved with water conservation can be made without a sacrifice in living standards, why do the inefficient methods of water use persist?

10.10 Managing Water Resources for Sustained Use

Concepts: economic efficiency, watershed management

The question at the end of the last section might be considered within the framework of innovation used in Chapter 2. Do we continue to use water inefficiently because alternative methods have not demonstrated a clear advantage? No. With respect to irrigation—the largest use of water worldwide, we know how to obtain higher efficiencies, but resistance to their adoption persists. Underlying the water efficiency problem is our economic perception of water and the reliance on technology to manage it (Vig. 10.1).

Vignette 10.1 The Perception of Water

Historically, water has been viewed by society--particularly by governments--primarily as an economic asset. In the MDCs large dams were built because they were the key element in the "multipurpose river basin development strategy" advocated during the period from 1910 to 1960. In this scheme, dams created reservoirs for flood control and recreation, and power was generated at the dam site.

economic efficiency

From the early years of water resource development, the need for flood control and navigation placed engineers in the forefront of water management. Engineers are technologically oriented, not ecosystem oriented (though this is gradually changing).

The coupling of the economic "water as value" view with the technology-based water management practiced by engineers created an "economic efficiency" approach to water resources management in the MDCs. Technology was often used as the sole source for providing solutions to water resource problems, particularly supply and power generation. This tendency is also becoming apparent in some LDCs, as ambitious dam building programs are afoot in Brazil, India, and China.

watershed management

What is absent from the economic efficiency approach is the consideration of water resources within the other elements of ecosystems. In this approach, watershed management is narrowly defined as the development of the economic potential of river basins, not as an integration of the biologic, atmospheric, and lithologic environments with the hydrosphere. We have already noted how elements which are linked within the hydrosphere, such as groundwater and surface water are still managed separately.

In order to obtain a durable supply of fresh water, we need to broaden the concept of watershed management to an ecosystem approach. Ecosystems include humans, which means our water management practices must evolve out of the dominant technological engineering and economic efficiency framework. A broader framework is needed which includes engineering, but uses a multifaceted approach to manage water as a renewable natural resource.

What other disciplines should be involved with ecosystem-based watershed management?

Applications / Discussion

10.2) Finish the table below:

Process	Pathways: Geosphere Components
Precipitation	atmosphere to lithosphere ; atmosphere to hydrosphere; ???
Runoff	lithosphere to hydrosphere ; hydrosphere to hydrosphere; ???
Groundwater Storage	???
Evapotranspiration	???
Evaporation	???

10.3) Larger water droplets have more kinetic energy when they strike the surface. How would surface erosion potential be affected by varying droplet size? What other factors related to precipitation would exert an influence over the erosion potential?

10.4) Many countries issue warnings for hurricanes, and evacuation is mandatory in some cases. Thunderstorms are more frequent and cause more deaths. How can the dangers from thunderstorms be minimized?

10.5) Examine the hydrographs below.

Which hydrograph is from an urban stream? How could you tell?
Why might an urban stream have lower baseflow than a rural stream?

10.6) In many urban areas, streams have been channelized, and in extreme cases their bottoms covered with concrete. In several countries efforts are currently underway to restore urban streams by removing the concrete, improving access to the stream, and limiting pollution inputs. What are the some of the possible benefits of these efforts? What are the drawbacks?

10.7) In what ways has the invisibility of groundwater affected our perception of this resource and its management?

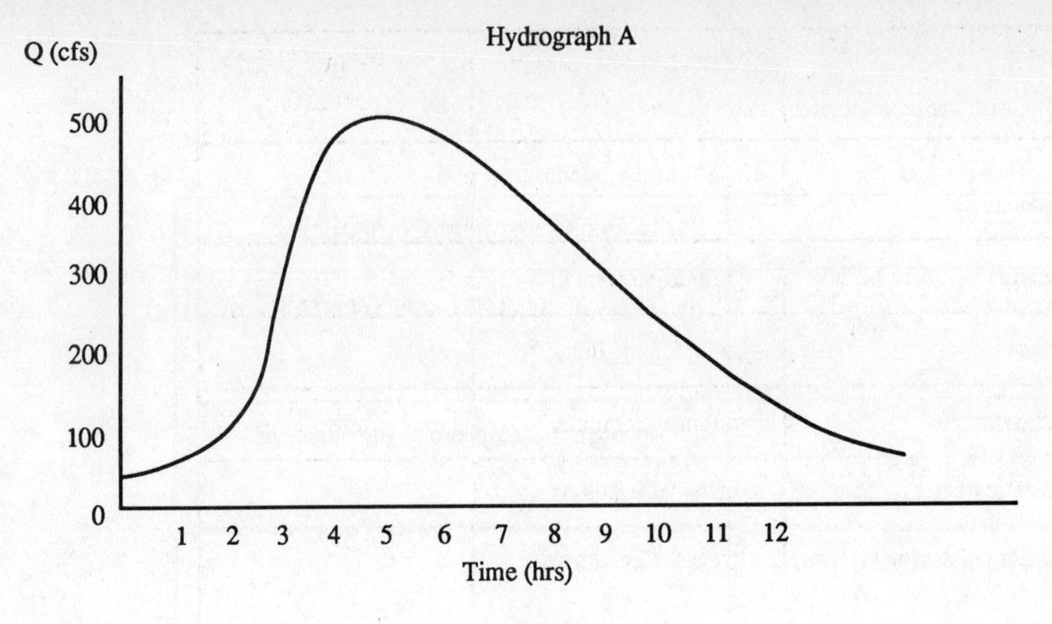

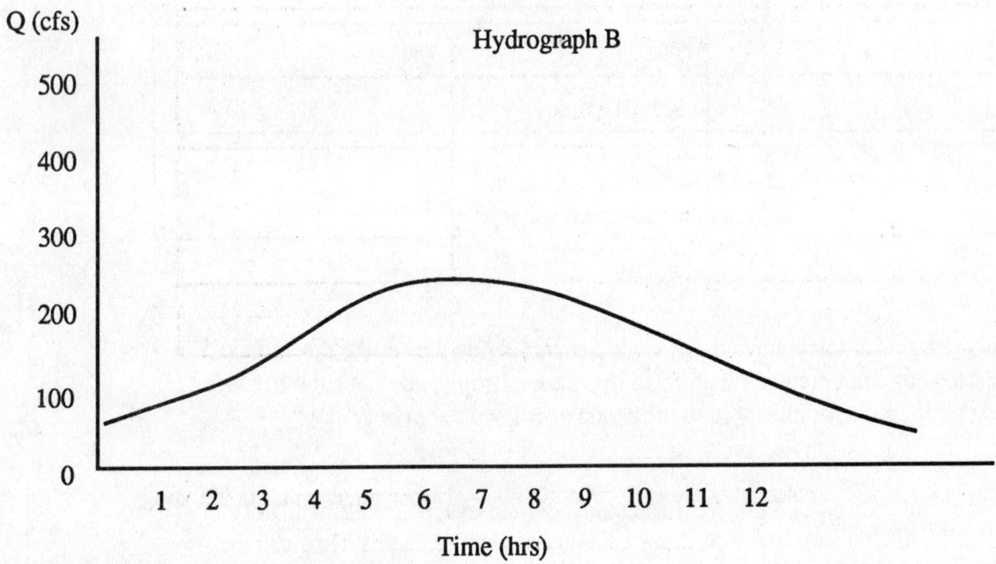

Figures for question 10.5

10.8) Identify the items in the water budgets of palustrine wetlands (swamps, marshes, bogs), and marine wetlands (tidal marshes). What is different? What is similar?

10.9) Assume it is summer, and evaluate your daily household water efficiency. Complete the table below:

Use	Quantity Calculation	Total Water Used (per day)
Lawn watering with garden hose	10 gallons per minute	
Showers	5 gallons per minute	
Baths	# taken * 34 gallons	
Bathroom sink	2 gallons per minute	
Washing dishes by hand	# meals * 20 gallons	
Washing dishes with dishwasher	# loads * 12 gallons	
Drinking	0.5 gallons per day	
Laundry	# loads * 40 gallons	
Toilet use	# flushes * 5.0 gallons if regular; 1.5 gallons if low-flush	
Total		

Where in your water use patterns is there a potential to save? How much water could you save with conservation?

10.10) Why can't water resource management problems be solved by technology alone?

Recommended Reading

Feldman, D.L. (1991). *Water Resources Management: in Search of an Environmental Ethic*. Johns Hopkins University Press, Baltimore, MD.

Kusler, J., and L. Larson. (1993). "Beyond the Ark: a New Approach to U.S. Floodplain Management," *Environment* (35), 6-11.

Lakeline (Journal of the North American Lake Management Society)

Sources of Additional Information

American Water Resources Association, 950 Herndon Pkwy. Suite 300, Herndon, VA. 22070-5528, (703) 904-1225

INTERNET: University Water Information Network.
Telnet address: gopher.c-wr.siu.edu

North American Lake Management Society, P.O. Box 5443, Madison, WI. 53705-5443, (608) 233-2836

United States Geological Survey, Water Resources Data (by state, by year)

References

Brooker, M.P. (1985). "The Ecological Effects of Channelization," *Geographical Journal* (151), 63-69.

Federal Interagency Floodplain Management Task Force, (1992). *Floodplain Management in the United States: An Assessment Report*. Natural Hazards Research and Applications Information Center, Boulder, CO.

Lvovich, M. (1973). *The World's Water*. Mir Publishers, Moscow.

Miller, A. and J.C. Thompson. (1975). *Elements of Meteorology* (2nd ed.). Charles E. Merrill, Columbus, OH.

Todd, D.K. (1980). *Groundwater Hydrology*. Wiley, NY.

United States Water Resources Council, (1978). "Floodplain Management: Guidelines for Implementing Executive Order 11988m," *Federal Register*, vol. 43, FR 6030-6055. February 10, 1978.

Viessman, W., and C. Welty. (1985). *Water Management: Technology and Institutions*. Harper and Row, New York, NY.

White, G.F. (1969). *Strategies of American Water Management*. University of Chicago, Chicago, IL.

11

WATER POLLUTION: PATTERNS, TRENDS, AND IMPACTS

Review

11.1 Summary

Water is everywhere, and so is pollution. Despite a seemingly renewable supply of water, shortages and public health problems persist in both LDCs and MDCs.

As the intensity of human activity increases, the vulnerability of water resources to pollution rises. Most agricultural and urban modifications to the landscape create residues transportable by flowing water. The oceans become the ultimate destination of this pollution, with the productive continental margins affected most.

Treatment systems for pollution in MDCs have been effective for limiting identifiable discharges from point sources such as sewer pipes. However, many streams and lakes remain polluted from the contributions of dispersed, or nonpoint sources. In LDCs, the high investment required for water treatment has stymied efforts to provide a reliable supply of drinking water.

Water resources are common property in most countries. As nations try to maximize the economic gains from short-term water use, the long-term consequences related to water quality are often ignored. Considerable effort will be required to develop a workable and equitable framework for global water management. Water pollution management strategies need to consider a wide range of the physical (such as land use development and stormwater runoff), and social processes (such as lawn pesticide applications), and their linkages within a watershed, while providing institutional oversight corresponding to the scale of these processes.

11.2 Types of Water Pollution

Concepts: classification, measurement, ranking, matrices

classification

Classification is a basic technique used by scientists to organize data. Once data are classified, they can be measured and ranked. Standards and criteria are required before measuring and ranking of the data groups can proceed.

To demonstrate these concepts, we can classify, measure, and rank water pollutants.

The criteria for classification are the pollutants' influences on the environment and human health. This is similar to the classification in *Environmental Geography* and yields the same eight classes of pollutants. We can then measure and rank the pollutant classes based on their impacts in different waterbodies.

measurement

Prior to ranking, the impacts, time frame, and study area—the measurement standards, are defined. Measurement of the impacts is based on the results of water quality samples. A low impact (coded as a "1") represents 1-5 samples exceeding regulatory limits for the pollutants. Medium impacts (coded as a "2") indicate 6-10 violations, and high impacts (coded as a "3") represent over 11 violations per year. No impact (0 violations) is coded as "0." For this example let's assume the sampling time frame is one month and the study area extent is a county and is called county "X."

ranking, matrices

Matrices can be used to perform the ranking. A sample matrix is shown in Table 11-1. In this case, the row elements are the pollutant classes, and the column elements are different waterbodies. The values in the cells are the relative environmental impacts of the pollutants present.

Greater impacts are represented by higher row totals. For county "X," the ranking of pollutant impacts is (rank in parentheses):

(1) tie-- Oxygen-demanding wastes; Disease-causing organisms; Toxic minerals and Inorganic compounds
(2) tie-- Synthetic organic compounds; Radioactive wastes
(3) Sediments
(4) Plant nutrients
(5) Thermal discharges

Here are some additional considerations for further analysis: What if the streams were sampled at different times than the lakes and wetlands? How might this affect the results? And, what advantage would there be to using the actual numbers of violations in the matrix cells instead of a code?

11.3 Measuring Water Pollution

Concepts: measurement, technology-based standards, water-quality standards, water quality indexes

In many nations, a widely held view until the 1960's was that large bodies of water could dilute the relatively small amounts of pollution discharged. The "solution to pollution is dilution," was a guiding principle of the early efforts to control water pollution.

There were two major problems with this approach: 1) pollution loads were increasing as growing cities and factories dumped their wastes freely into lakes and streams; and, 2) low concentrations of certain pollutants, such as DDT generated significant environmental risks.

As science and society observed water pollution spread, the traditional views and approaches to water pollution measurement and control changed. The events spurring this were dramatic: rivers caught fire, such as the Cuyahoga in Ohio, and Lake Erie became

TABLE 11-1 Relative Impacts on Waterbodies in County "X"
Relative impacts: 0 =no impact (0 violations), 1= low (1-5 violations), 2=medium (6-10 violations), 3=high (>11 violations)

Pollutant Class	Impacts			
	Wetlands	Lakes	Streams	Total
Oxygen demanding wastes	3	3	1	7
Plant nutrients	2	2	0	4
Sediments	3	1	1	5
Disease causing organisms	3	3	1	7
Toxic minerals and inorganic compounds	3	3	1	7
Synthetic organic compounds	3	2	1	6
Radioactive wastes	3	2	1	6
Thermal discharges	1	0	0	1

eutrophic.

At a less obvious level, bioaccumulation, secondary chemical reactions, and the persistent toxicity of some pesticides and industrial discharges required a finer measurement of the wastes entering our surface and groundwaters.

measurement

Measurement technology improved. Refinements to some measurement devices, e.g., the spectrophotometer, improved the measurement precision of pollutants to the part per billion. The most common measurement of pollutant concentration now being used is milligrams per liter (mg/l). This is equivalent to parts per million (ppm).

Two key issues of pollution control evolved from the improvement in measurement technology: What are the safe limits of pollution; and, how do we achieve those limits? (Vig. 11.1).

Vignette 11.1 Pollution Control Standards

In order to achieve acceptable levels (what constitutes "acceptable" varies by nation) of pollution, many nations adopted two standards of pollution control: technology-based standards, and water-quality based standards.

technology-based standards

Technology-based standards reflect the best level of pollution control attainable from the best available technology. Sources of pollution discharges (factories, wastewater treatment plants), were required to incorporate the latest technology in their treatment before releasing processed water to the receiving waterbodies. Currently, the best level of pollution control available for wastewater is a three stage treatment process consisting of settling, decomposition of organic wastes, and filtration and disinfection.

water-quality standards

Under the water-quality based approach, a receiving waterbody was designated a certain use it could support, such as drinking water supply, body contact, recreation, or irrigation. The allowable amounts of pollution discharged were scaled to the type of designated use.

Drinking water is the "highest use," as it requires the cleanest water. More extensive treatment is therefore required before discharging effluent into a stream or lake with this designated use.

Problems exist with both approaches. Many smaller communities lack the financial resources to implement advanced water treatment under a technology-based approach. On the other hand, some communities might be "let off the hook" for pollution control if the nearby waterbody is assigned a designated use which does not require as much treatment.

How can environmental equity (fairness) be achieved in water quality preservation--is there a way to divide the burden of regulation and monitoring equally among communities of different size and financial capacity?

Some critics of the approaches cited above advocate a uniform standard for measuring water quality using an index of several chemical and physical indicators. A Water Quality Index (created by the National Sanitation Foundation) is shown in Table 11-2.

Parameters used in an index are typically considered to be the key chemical and physical indicators of water quality. After testing for each parameter with specialized equipment, such as a dissolved oxygen meter, the results are recorded and transferred to a weighting curve where a numerical "Q-value" is obtained (Fig 11.1). The Q-value is then

TABLE 11-2 Sample Water Quality Index

Parameter	Q-value	Weighting Factor	Total
Dissolved Oxygen	97	0.17	16.49
Fecal Coliform	19	0.16	3.04
pH	85	0.11	9.35
Biochemical Oxygen Demand	51	0.11	5.61
Temperature	92	0.11	10.12
Total Phosphate	97	0.10	9.7
Nitrates	98	0.10	9.8
Turbidity	76	0.08	6.08
Total Solids	20	0.06	1.20
	Overall Water Quality Index = 71.39		

multiplied by a weighting factor. Dissolved oxygen has a relatively high weighting factor, and thus is more significant in determining water quality than the other parameters.

The sum of the parameter values are added to compute the Water Quality Index (WQI). A WQI rating of 50-58 would meet most states' water quality standards. However, a waterbody with a high index value could still be unsafe for some uses, such as swimming. For instance, if the eight other parameters looked good, but fecal coliform was present in high concentrations (greater than 200 colonies per 100 ml), body contact would be prohibited by health officials.

The index shown here can be extended to include consideration of a waterbody's physical condition and environment: e.g., water levels, bank erosion, and adjacent vegetation. However, the equity issue remains. Without adequate funding for pollution control, a uniform measurement standard does not remove the disproportionate financial burden to communities with differing financial capacities.

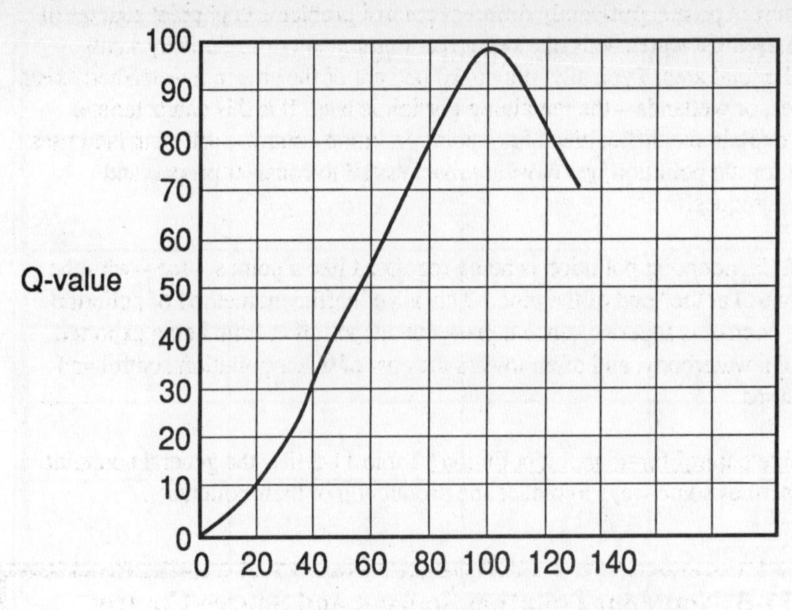

Figure 11.1 Dissolved oxygen Q-value curve

11.4 Water Pollution Sources and Distribution Processes

Concepts: point sources, nonpoint sources, classification, land and water management, source control

point sources

There are two broad types of water pollution sources: point and nonpoint. Point sources of water pollution originate from a known discharge point, usually a pipe or ditch. Municipal sewage treatment plants, industry, and power plants are the principal point source emitters of water pollution.

nonpoint sources

Nonpoint water pollution is spatially dispersed, comes from many locations, and in most cases is instigated by rainfall or snowmelt. Stormwater runoff from urban and agricultural areas, and acid precipitation are the primary forms of nonpoint pollution. Groundwater pollution may have nonpoint origins, such as the seepage of pesticides from irrigation. This is discussed in the next section.

classification

The classification of water pollution sources into point and nonpoint origins offers opportunities for further analysis. For instance, in the United States most of the point sources, which are easily traceable, have been regulated. This means that the discharge of specific pollutants from each site is known. Given this knowledge, we can sample the water quality of a stream and obtain the pollution contributions from nonpoint sources.

Despite a comprehensive point source regulatory program in the U.S., substantial pollution contributions from nonpoint sources persist. As of 1995, nonpoint pollution is still preventing many streams from becoming "fishable and swimmable"—a primary goal of the Clean Water Act of 1977.

Nonpoint sources pose significantly different control problems than point sources of water pollution. In most watersheds, the water features constitute a relatively small percentage of the total area. Typically, only 5-10 percent of the area in a watershed exists as streams, lakes, or wetlands—the remaining portion is land. It is this characteristic which helps to explain the difficulties of nonpoint pollution control—differing land uses are responsible for the pollution, yet most approaches fail to consider proper land management techniques.

land and water management

In many MDCs, nonpoint pollution is being regulated like a point source—with the emphasis on control at the "end of the pipe." This is an inefficient method of pollution control. Source control is superior, since it prevents the pollution from being exported and delivered to a waterbody, and often lowers the cost of water pollution control and treatment measures.

What is source control for nonpoint pollution? Table 11-3 lists the general nonpoint sources and describes some ways to reduce the production of their pollutants.

source control

TABLE 11-3 Nonpoint Pollution Sources and Source Control

Nonpoint Source	Source Control Measures
Urban Stormwater	reduce erosion, a large percentage of stormwater contaminants are concentrated in the soil; use basic soil erosion methods in urban areas including construction site stabilization with seeding, vegetation, and mulches, slope stabilization, and detention ponds
Agricultural Runoff	irrigation techniques can be improved so less water is applied and less runoff occurs; the general sustainable methods of landscape maintenance discussed in Chapter 6 can be used
Acid Rain	short term: use low-sulfur coal, or substitute a non-fossil fuel; long-term: a decentralization plan for industry and dispersal of the urban heat island can help

At smaller scales, techniques also exist for control of nonpoint pollution. For individual lots in urban areas: 1) lawn fertilizer use can be reduced; 2) leaves left on the lawn slow the runoff into storm drains, thus allowing for more infiltration of rainwater into the soil; 3) lot grading for new construction can channel water into depression storage on the lot, instead of speeding it into a storm drain; and, 4) water storage devices, such as cisterns can be used to save rainwater for heating, washing, and lawn watering.

11.5 Groundwater Pollution

Concepts: vulnerability, land and water management

vulnerability

Whenever pollution is produced on the surface, the threat to groundwater increases. The same principles of source reduction discussed in the previous section apply to groundwater pollution. Proper management of surface land uses is critical to lowering the vulnerability of groundwater to pollution.

land and water managment

Most land uses contribute to groundwater pollution, which can have point or nonpoint origins. As a point source, hazardous waste spills and dumping from localized industrial and military sites often pollute groundwater. When considered as a nonpoint source, groundwater pollution occurs as nutrient-rich irrigation runoff from agricultural areas spanning several counties or other political boundaries. Elevated nitrate levels have been detected in several states overlying the large Ogallalah aquifer in the central United States.

Examination of a specific type of groundwater pollution lends some insights into the general vulnerability of groundwater to pollution produced by surface activities. Table 11-4 shows the features of septic tank systems which contribute to the threat of groundwater pollution.

TABLE 11-4 Features of Septic Tank Systems Contributing to Groundwater Vulnerability

Feature	Possible Effect
Improper Construction	possible leakage of effluent into groundwater
Siting	soil is incapable of drainage
Maintenance	inadequate maintenance can cause septic tanks to overflow
Depth to Water	septic drainage is more likely to reach groundwater if the water table is higher
Density	too many septic fields in an area can expose groundwater to higher levels of pollution

Density is the key feature determining the risk septic tanks pose to groundwater. When the density of septic tanks increases, the soil's capacity to purify the septage may be overcome.

These features are common to other types of land uses and also contribute to their potential for generating groundwater pollution. In many countries, the improper siting of landfills has created wide areas of contaminated groundwater, while poorly maintained sanitary sewer lines often leak and infiltrate into groundwater.

11.6 Pathways and Sinks in the Hydrologic Cycle

Concept: residence time

After pollutants are produced, they ultimately end up in sinks. Pollution sinks are areas where the accumulation of pollutants exceeds the natural flushing or purification actions of the receiving media. Soil is a sink for organic pollutants and heavy metals. Inland lakes are sinks for the pollution contained in stormwater runoff, and the oceans are the ultimate sink for all the pollution carried by flowing waters.

residence time

Once pollution enters a sink, it stays resident for a certain period of time. The length of time a pollutant stays in a sink is called its residence time, and this time period is dependent on the mobility of the receiving medium and outside events which may disturb the medium. Pollutant composition also affects the residence time. Some pollutants, such as sewage, may decay or biodegrade in a few days; others, like toxic metals may persist for years. These persistent pollutants are called "conservative."

Soil is relatively immobile and will hold on to pollutants for very long periods unless disturbed by plowing or construction activity. Subsequent precipitation or wind—both outside events--can then erode the soil and transport the pollutants to another sink. Conversely, rivers are short-term sinks of pollution because their flows are continually replenished from groundwater and overland stormwater runoff.

As the above examples show, pollution is often transferred from one sink to another. Bioaccumulation is an example of how different sinks (trophic levels) in a foodchain pass and accumulate pollutants. The residence time in a foodchain is longest at the highest trophic level. This increases the pollution risk to humans and other omnivores.

From a spatial perspective, pollution sinks are often associated with areas of high bio-diversity and productivity. Shorelines and continental shelves are prone to large amounts of runoff from land activity, boat engine pollution, and deposition of river sediment loads containing pollutants. In wetlands, the balance of inflows and outflows maintains consistent water levels, but this same balance slows the exchange of water and creates longer residence times for some pollutants, such as road oil.

11.7 Pollution of the Oceans

Concept: common resources

In the last section, the oceans were mentioned as the ultimate sink for pollution in flowing waters. Other forms of water-related pollution end up in the oceans; e.g., acid rain also contributes pollution. Because of the vastness of the oceans, we don't often think of the threat to them posed by pollution, but the impacts are real. High bacteria levels have caused beach closings or advisories along sections of the U.S. coast (Table 11-5).

common resources

As shown, the oceans are not immune to pollution. Since the oceans are common resources and nobody owns them, there is a tendency to over-exploit the resource. There is no direct charge to the polluter when wastes are dumped into the ocean; the cost is external and borne by society as a whole.

The problems associated with the common ownership of resources was addressed in a

TABLE 11-5 Beach Closures or Advisories Due to High Bacteria Counts (days)

State	1989	1990
Maine	1	72
Massachusetts	60	59
Rhode Island	0	0
Connecticut	103	218
New York	923	581
New Jersey	266	228
Delaware	62	11
Maryland	0	0
Florida	N/A	234
California	338	64
Totals	1,753	1,467

Chart courtesy of the National Resources Defense Council (NRDC).
Source: Kassalow. (1991).

famous essay by Garrett Hardin entitled "The Tragedy of the Commons." Originally, "the commons" was an area of pastureland in England provided free for use by anybody who wanted to graze cattle. There were no problems on the commons while the number of animals was relatively small compared to the total pasture area. But each herder tried to maximize their short-term gain by grazing as many cattle as possible; if their cattle didn't eat the grass, somebody else's would. This resulted in overgrazing of the commons, depletion of the resource, and everybody lost their herds as the animals died of starvation.

What does this event imply for environmental regulation?

11.8 Pollution Control with Treatment Systems

Concepts: hierarchy of water uses, graywater, conjunctive use, source control

Treatment systems for human sewage can be classified into two major categories: high-tech, and low-tech. The high tech approach was outlined in section 11.3, and consists of wastewater treatment plants employing a process of settling, decomposition of organic wastes, and filtration and disinfection.

In high-tech systems the final destination for the treated liquid wastewater is usually a stream; solids are dewatered and sometimes incinerated. Some areas now permit the application of the dried solids, called sludge or biosolids, to agricultural fields.

Wastewater treatment plants require heavy capital investments for the basins used in settling (clarifiers) and organic decomposition. Significant expense is also required for filtration, since large buildings (and land areas) are required to provide room for the sand or mixed media filters using sand, garnet sand, and granular activated carbon.

hierarchy of water use

One disadvantage of high-tech systems is the large quantity of clean water used to process the waste. Why not use processed wastewater to process waste? This concept recognizes the hierarchy of water uses; it matches the quality of water used with the quality of the task. It is intuitively similar to energy efficiency using the Second Law of Thermodynamics, where the quality of energy is matched to the task.

graywater

conjunctive use

Processed wastewater is often called graywater. It is not fit for drinking, but is well suited to lawn watering and irrigation—two major consumptive uses in urban and rural areas. This provides an opportunity for a form of conjunctive use between urban and rural areas. Treated urban wastewater can be used for irrigation, and excess agricultural irrigation runoff can be captured for urban lawn watering.

Low-tech approaches to sewage treatment rely primarily on soil absorption. There is little use of chemicals or machinery in these systems. At the house level, soil is used as a filter medium in a septic system. Larger scale low-tech treatment systems involve three steps: 1) settling and filtering; 2) decomposition of effluent in ponds by bacteria and algae; and, 3) application of the treated water to fish ponds, croplands, or wetlands.

At larger scales, low-tech wastewater treatment approaches presents two general problems: 1) there is a need to load the ponds conservatively with waste, otherwise the aerobic decomposition process will become anaerobic (without oxygen), and severe odors will result; and, 2) groundwater contamination is possible, so the bottoms of the ponds should be sealed to prevent infiltration of wastewater into the water table.

source control

Potential benefits from source reduction are available to the low and high-tech approaches. Residual solid wastes pose a significant problem in both systems, as they often contain high concentrations of toxic metals. Source reduction practices can provide the benefit of preventing the addition of metals to the waste stream, thus avoiding the problems of disposal encountered later.

11.9 Summary on Water Quality: How Well Are We Doing?

Concepts: watershed management, scale

Despite heavy investments in technology, water quality problems persist in MDCs. The control of point source pollution has been effective, but nonpoint source pollution in urban and agricultural areas continues to degrade stream and lake water quality.

In LDCs, uncontrolled urban growth and deforestation increase the erosion potential, while the application of toxic and persistent pesticides continues. Obtaining fresh water in many LDCs is still problematic, and waterborne diseases present major health problems.

The responses to water pollution have been generally fragmented. Basic recognition of the hydrologic and environmental linkages between groundwater and surface water, quantity and quality, and simultaneous land and water management has been lacking. And, most pollution control laws address problems after they are within the water cycle, rather than preventing their entry.

watershed management

There is the need for a recognition of watersheds as more than physical divisions between drainage networks. An integrated approach to watershed management must be ecosystem-based, and account for the interaction of the social and physical environments within watersheds.

A complete concept of watershed management uses not only the site, but the situation of the area. Within each watershed (its site), there are multiple layers. Flows and energy exchanges occur within and between the atmosphere, surface, and groundwater components. Management and pollution control strategies must account for these linkages. For instance, groundwater quality is highly dependent on land use activities at the surface.

Considering the situation of watersheds, each watershed is influenced by external events—with impacts across different media. Acid precipitation produced in one watershed often affects the surface and ground water quality of another watershed.

scale

These problems illustrate why institutional frameworks for watershed management and pollution control need to operate at the correct scale. While the current trend in many MDCs is to offload pollution control to local jurisdictions, the problem requires institutional attention at the scale encompassing the physical and social processes at work. This scale is often national, and in many cases international.

Applications / Discussion

11.2) Outline a sampling program for the water quality in the streams, wetlands, and lakes within your local watershed. Specify a manageable number of water quality parameters (3 or 4). The objective is to obtain comparable results for the environmental impacts on each. Some things you should consider are the timing of the sampling, and the location of the sampling (distance from the shore).

11.3) Perform several shoreline "grab" samples in a local stream and construct its water quality profile. A simple jar can be used for a grab sample. Analysis equipment is usually available from chemistry or biology departments. If no equipment is available, you can have a local EPA-certified lab analyze the samples for a nominal cost. Try to obtain results for the parameters in Table 11-2.

11.4) Make a list of the practices which can reduce urban stormwater pollution at its source.

11.5) Obtain a land use map for your city or local region (the study area) and construct a map which shows the different types of threats to groundwater. Create 4-7 general categories (eg., chemical spills, septic tank leakage, etc.) of groundwater threat. Base your assessment on the intensity and character of the surface land use zones. For instance, a dense residential area which has sewers could threaten groundwater if the pipes leaked (a "pipe leakage" threat), whereas a less dense residential neighborhood with septic tanks would pose a differnent type of threat to groundwater ("septic failure"). Your final product is a map identifying the areas where the different threats to groundwater you defined exist.

11.6) Which is less harmful: a conservative pollutant in a river, or a non-conservative pollutant in a lake? Make a list of the information you would need to answer this question.

11.7) How are the controversies over territorial limits (Chapter 2) and ocean pollution related?

11.8) At the household scale, analyze the hierarchy of water uses. Under the column heading "Minimum Water Quality Required," fill in graywater or clean water. Complete the outline below:

Use	Minimum Water Quality Required
Lawn Watering	graywater
Showers	clean water
Baths	
Brushing Teeth	
Washing dishes (machine)	
Washing dishes (manual)	
Cooking/Drinking	
Laundry	
Toilets	

At the household level, is the amount of clean water conservable by using graywater a significant percentage of the total? Can graywater be incorporated into single-family homes without a complete redesign of the plumbing? How?

11.9) Since point sources are the most concentrated forms of pollution--and we have controlled these well--pollution is not the threat it once was to the world's waters. Discuss this statement.

Recommended Reading

Hardin, G. (1968). "The Tragedy of the Commons," *Science* (162), 1243-1248.

Hynes, H.B.N. (1974). *The Biology of Polluted Waters.* Univ. of Toronto Press, Toronto, Ontario.

Journal of Soil and Water Conservation (monthly), published by the Soil and Water Conservation Society (address below)

Yeager, P. (1991). *The Limits of the Law: the Public Regulation of Private Pollution.* Cambridge Univ. Press, Cambridge, England.

Sources of Additional Information

Center for Marine Conservation, 1725 DeSales St. NW, Suite 500, Washington, D.C. 20036, (202) 429-5609.

Clean Water Action, 317 Pennsylvania Ave SE, Washington, D.C. 20003

INTERNET: Surface Water Pollution Gopher on the North American Menu

Soil and Water Conservation Society, 7515 N.E. Ankeny Rd., Ankeny, IA. 50021-9764, (515) 289-2331.

References

American Public Health Association. (1990). *Standard Methods for the Examination of Water and Wastewater* (16th ed.). American Public Health Association, NY.

Kassalow, J., et al. (1991). *Testing the Waters: A Study of Beach Closings in Ten Coastal States.* Natural Resources Defense Council.

McGhee, T.J. (1991). *Water Supply and Sewerage* (6th ed.). McGraw-Hill, NY.

Whipple, W., Jr. and J.V. Hunter. (1977). "Nonpoint Sources and Planning for Water Pollution Control," *Journal of the Water Pollution Control Federation,* (49), 15-23.

Yates, M. (1985). "Septic Tank Density and Ground-water Contamination," *Ground Water* (23), 586-591.

12

HAZARDOUS WASTE PRODUCTION AND DISPOSAL

Review

12.1 Summary

It would be a better world without waste, but we are stuck with it. Increasing consumption of goods in MDCs, and population growth within LDCs are both operating to increase the amount of waste produced. The problem is, not only is all waste increasing, some of it is dangerous.

Hazardous waste represents the residues of substances harmful to human health and other organisms, with risks incurred at all phases of the product life cycle—distribution, use, and disposal. Quite often, the effects of these substances remain dormant for many years after their initial use, with much uncertainty and debate over the safe limits of their use.

Various methods have been developed for the reduction of the risk generated by hazardous waste. Protective clothing for workers and stronger storage containers are used at the production and distribution life cycle phases. Most of the concerns for reducing risk, however, are now focused on the vulnerability of disposal sites and the release of toxins during treatment. Here, physical processes may instigate releases of toxic substances into the environment.

Since both waste disposal and treatment have potential side effects, waste reduction at the source and recycling have emerged as effective ways to reduce the volume and level of risk associated with hazardous wastes. As part of a program of integrated waste management, these techniques can help to improve the performance of disposal and treatment by reducing the impacts on ecosystems.

12.2 Types and Sources of Disposable Waste

Concepts: risk, product life cycle

Hazardous waste is waste which poses a threat to human health and the environment. Most hazardous waste is generated by three sources: chemical manufacturing, primary metal manufacturing, and petroleum refining. If classified by waste composition, hazardous wastes include heavy metals (lead, zinc, and arsenic), synthetic organic compounds (PCB, DDT, and dioxin), petroleum products (grease, oil, and gasoline), acids (hydrochloric and sulfuric acid), biological substances (bacteria and plant toxins), and radioactive materials (nuclear fuel rods and nuclear medical materials).

risk

It is during their transport when hazardous wastes pose the greatest risks to humans and the environment. The risk from hazardous substances is related to the likelihood of adverse effects resulting from exposure or proximity to the substances—temporal and spatial factors. One method of risk reduction used by many communities has been to route trucks carrying hazardous materials away from densely populated areas. Table 12-1 shows the temporal and spatial characteristics of DDT exposure.

TABLE 12-1 Temporal and Spatial Characteristics of DDT Exposure

Life Cycle Event	Frequency of Exposure	Areal Extent	Concentration within Affected Area
Production	low	concentrated	high
Distribution	low	wide	high
Use	high	diffuse	low/moderate
Secondary Dispersion	high	diffuse	low

product life cycle

The primary temporal (frequency of exposure), and spatial (areal concentration and density within area) characteristics of exposure vary throughout the hazardous waste product life cycle. The product life cycle for hazardous substances includes production, distribution, use, and its secondary dispersion, which requires further definition.

For most substances, the secondary dispersion is influenced by climate, chemical composition of the substance, and the disposal/treatment method of the waste residues. For instance, prevailing winds may transport airborne substances long distances, chlorine added to hydrocarbons increases their persistence in the environment, and the location and/or efficiency of incineration, landfilling, or dumping can affect the exposure to the substance.

In the table, the environmental and human risks from exposure to DDT are greatest during its use and secondary dispersion. Regarding its use, DDT was widely sprayed to control pests in sparsely populated agricultural areas, and sprayed in many highly populated urban areas to combat Dutch Elm disease. As to DDT's secondary dispersion, winds continue to deposit DDT in the Great Lakes, although its use has been banned for over 20 years in the U.S.

Different hazardous substances will exhibit different characteristics. This table provides a practical way to identify the potential risks associated with exposure to a hazardous substance. Unfortunately, many hazardous chemicals were used and disposed of improperly before we understood their full effects on human health and the environment.

What segment(s) of society (private industry, the government, the public), should bear the clean-up costs for hazardous substances which have been banned, but still persist in the environment?

12.3 Traditional Methods of Waste Disposal

Concept: siting

Waste disposal methods have changed in response to urbanization and technological improvements. During the early growth of cities accompanying the Industrial Revolution, open dumps on the outskirts of cities were used for solid waste. Over time, science and technology created new substances which rendered traditional disposal methods ineffective. For instance, large barrels of hazardous waste brought into dumps would rust and leak into groundwater. In response, new methods of disposal were adopted, including sanitary landfills designed to prevent the penetration of leachate into groundwater. Meanwhile, increasing consumption in the MDCs was generating higher amounts of waste, creating severe shortages of available land for waste disposal. Some cities, such as Philadelphia, ran out of landfill space.

In the United States, hazardous waste production was also increasing, as the per person amount rose to one ton a year. Most hazardous wastes were placed in landfills, and these sites still contain the majority of hazardous materials. However, as subsequent events have shown, the quantity and toxicity of hazardous wastes required a more careful selection of their final location.

siting

The decision process taken to locate a facility or activity is called siting. Initially, the siting criteria used for most landfills required the easy access for waste haulers, and level, vacant land. As a result, waste dumps were often located outside, but close to the current urban boundaries, and sometimes in floodplains or areas with hydric soil resulting from high water tables.

As a land use, waste disposal must compete with other and more desirable land uses for space. Social concerns over safety and the removal of land which may produce more economic returns are now greatly influencing siting decisions. What considerations can help a community make the decision between allocating land for waste disposal, or developing that same land for a new subdivision?

12.4 Some Consequences of Improper Waste Management

Concept: science and the law

As urban areas expanded, many former waste dumps were filled in and overlain by development. The tragedy of Love Canal in Niagara, New York represents the worst consequences of poor siting: proximity to population, and a failure to account for the properties of soil moisture on waste over time.

science and the law

The delayed release and effects of toxics on humans and the environment raises many questions for science and law. A key question within this context is: What are the needs and consequences of regulation now, given imperfect knowledge about the effects and the possibility future conditions will change? The debate over chlorine compounds illustrates these issues (Vig. 12.1).

Vignette 12.1 Should Chlorine Compounds be Banned?

The following is excerpted from a debate between Chlorine Chemistry Council managing director Brad Lienhart and Greenpeace toxicology researcher Joe Thornton.

Opening statements: ecology and health

Thornton: Organochlorines are known to disrupt the body's hormones, to cause genetic mutations and metabolic changes, to cause or promote cancer, infertility, impair childhood development, to cause neurological damage, and to damage the immune system.

Here in the Great Lakes [region], scientists have documented epidemics among virtually every predator species. In each case, the consumption of Great Lakes fish contaminated with organochlorines appeared to be the cause. The available science suggests that the effects are beginning to occur already among people.

We need to treat organochlorines as a class. There are 11,000 in commerce plus thousands more that are produced as by-products. It would take centuries to regulate these one-by-one. Those that have been tested have overwhelmingly turned out to [demonstrate] persistence, toxicity, or bioaccumulation. There's good reason for this. When we add chlorine to hydrocarbon, that new organochlorine is almost always more chemically stable, more resistant to breakdown, tends to be more bioaccumulative, and it also tends to increase the toxicity of the compound.

It makes sense to treat organochlorines as guilty until proven innocent. With one action, we can [cease production of] the largest group of persistent toxic substances in the global environment.

Lienhart: The stakes are high in this discussion. We must balance the risks of chlorine chemistry against not having access to chlorine and look carefully at the alternatives where they do exist because they may have adverse impacts of their own. [We have] to be sure that we are doing what is right for the ecosystem as well as the economy. [Joe] offered what seems to be a pretty easy answer. [But] if we act wrongly or prematurely we could hurt the very natural resources that we're trying to protect. [The process] must be fair and balanced.

The products have distinct chemical and physical characteristics, and you cannot generalize about this chemistry. Chlorine's role in society includes many public and human health advances such as water disinfection, pharmaceuticals, and medical technology, such as all the x-ray processes and film manufactured in the world. The World Health Organization says that the [50%] extended lifespan since the 1930s can most significantly be attributed to chlorine chemistry.

In many cases chlorine is used in place of products that are flammable or pose another very real or very immediate health hazard. DDT has been replaced by pesticides that are not persistent, toxic, or bioaccumulative, but they're still made out of chlorine chemistry. The Charles River Association study conservatively says the benefit [of chlorine chemistry] is $100 billion a year.

Rebuttal: economy and alternatives

Thornton: A chlorine-free economy will be a healthier economy. Totally chlorine-free process would save the paper industry $185 million-$370 million in chemical costs, $108 million-$189 million in saved energy costs, and millions or billions for improved worker productivity, lower costs for waste disposal, and lowered liability.

PVC should be an obvious one, traditional materials and chlorine-free plastics can replace PVC for virtually all uses. In Europe, dozens of communities, manufacturers, and hospitals have virtually replaced PVC in a few years.

A National Academy of Sciences found that [alternative farming] methods are effective, save the farmers money, and improve financial returns.

Lienhart: This is a free-market economy. Certainly people are free to choose alternatives to chlorine chemistry, if they exist, if they're practical. The reason we have 15,000 different applications of chlorine chemistry is that chlorine chemistry is advancing substitution chemistry and it brings value to the free marketplace. We have more than 300,000 studies in place that reflect on chlorine chemistry. An interpretive overview of the potential effects of chlorine effects on human health and the environment is being completed.

Where the alternatives do exist, we must look at the [costs and benefits] through the same process parameters. there are by-products of ozonation that are potentially very toxic. You can take petrochemicals and pesticides out of the farm environment. But when you put alternatives in place you have to increase the crop land. You must do a proper input-output analysis to make sure the alternatives do not do greater harm that the pesticides you are concerned about.

We need to work through a sound scientific process.

Closing statements: the next step

Thornton: How much science do we need before we can take action? If our approach is to consider chemicals innocent until proven guilty, the we don't have enough proof to phase out chlorine. But if our priority is preventing disease, preserving the environment, preserving human health, then we do know enough now.

Lienhart: [The] process should include scientific integrity, all of the stakeholders at the table, and a clear understanding of the socioeconomic values. The application and the alternatives should be looked at through the same process parameters. We'll be at the table to participate.

Much of the debate revolves around the issues of costs, benefits, and risks associated with the continued use of chlorine compounds. Does "a sound scientific process" need to focus on the costs and benefits of a substance, or place its main emphasis on the risk to the natural environment and human health?

12.5 Hazardous Waste Regulation in the United States

Concepts: adaptive policy, sustainability

adaptive policy

The formation of environmental policy is an adaptive process. As the debate over chlorine demonstrates, there are many groups with differing views on issues. The current debate over the reauthorization and modification of the major legislation to clean up toxic sites—"Superfund," is another example of how diverse views compete in the policy arena. In trying to decide who should pay the bill for cleaning up toxic sites, many different groups are proposing different approaches.

Four considerations have traditionally predominated the environmental policy arena in the United States: economic development, conservation, public health, and the preservation of features such as natural parks and wilderness. From these have emerged different beliefs, or ideologies about how environmental policy should be constructed. On most issues, the policy is adaptive—it is the product of compromise and conflict between the different beliefs and interest groups.

Much of the debate is related to the broad question of how natural resources should be managed: Do we expand capacity to meet growing demand (development), or influence consumption/use with an incentive/penalty system, e.g., Polluter Pay Laws (regulation)? These options are shown in Figure 12.1.

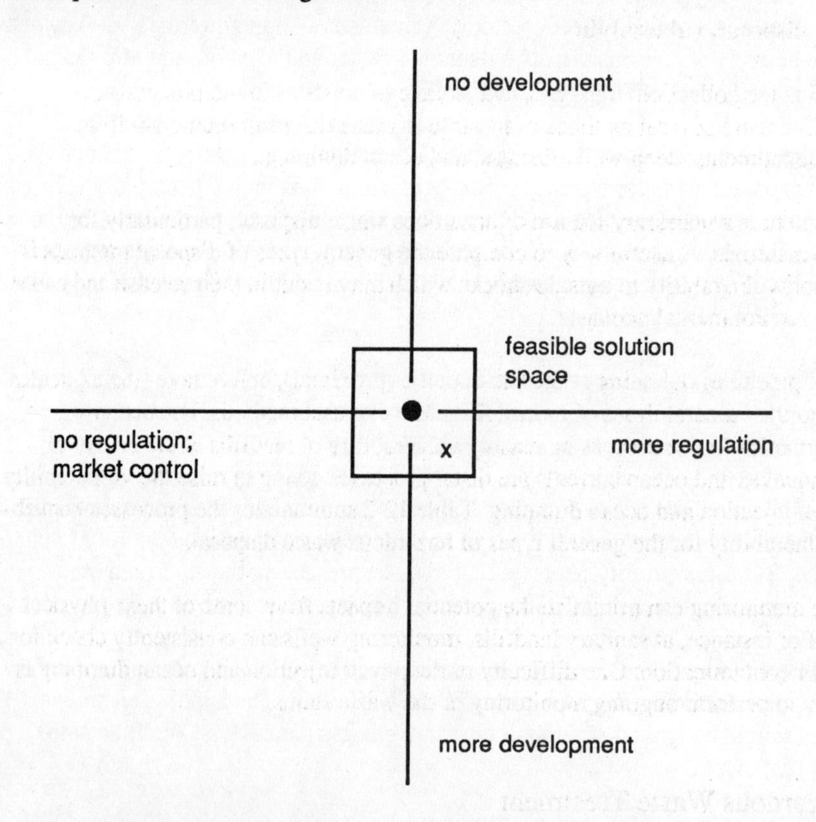

Figure 12.1 Adaptive environmental policy process

The feasible solution space defines the general boundaries for development and regulation—with limits depicted for both. Population growth and economic expansion require some expansion of capacity (development), while at the same time this places increasing stress on the environment, so some regulations are required.

Society's values determine the current policy within the feasible solution space. Throughout U.S. history, the approaches to environmental management and resource use have focused on economic/quantitative issues to meet relatively short term industrial and agricultural needs. The "x" represents this tendency; development has initially been favored, and regulations have followed after the environmental problems emerged.

sustainability

The feasible space implies that economic development and environmental protection are not mutually exclusive. This is a necessary ingredient for achieving sustainable resource use. The "o" in the diagram represents the balance between economic development and regulation.

The diagram is a broad generalization, but it conveys the challenge of achieving sustainable policies within the context of competing and conflicting interests.

12.6 Modern Hazardous Waste Disposal Methods

Concepts: disposal, vulnerability

disposal

Disposal is the collection, transport, and storage of wastes with no processing or treatment. There are several methods of hazardous waste disposal: secure landfills, surface impoundments, deep-well injection, and ocean dumping.

vulnerability

Confinement is a necessary feature of hazardous waste disposal, particularly for radioactive materials. A useful way to compare the general types of disposal methods is to assess their vulnerability to outside shocks which may result in their release and cause subsequent environmental damage.

Physical processes operating at the site-specific (proximal), and remote (distal) scales contribute to the vulnerabilities of hazardous waste disposal methods. For instance, leachate formation is one process increasing vulnerability of landfills at the proximal scale; earthquakes and ocean currents are distal processes acting to raise the vulnerability of deep-well injection and ocean dumping. Table 12-2 summarizes the processes contributing to vulnerability for the general types of hazardous waste disposal.

On-site monitoring can minimize the potential impacts from some of these physical processes. For instance, at sanitary landfills, monitoring wells can consistently check for groundwater contamination. One difficulty of deep-well injection and ocean dumping is the inability to perform ongoing monitoring of the waste status.

12.7 Hazardous Waste Treatment

Concept: treatment

treatment

Treatment of hazardous wastes involves the processing of the waste to make it less harmful. Because of increasing health concerns and regulations, the current trend is away from land disposal, and toward on-site treatment.

TABLE 12-2 Physical Processes Contributing to Waste Disposal Vulnerabilities at Different Scales

Disposal Method	Proximate	Distal
Secure Landfill	leachate formation, precipitation, erosion, runoff	groundwater flow
Surface Impoundment	leachate formation, evaporation, precipitation, erosion, runoff	groundwater flow, wind patterns
Deep-well Injection	bedrock fracture	earthquakes, groundwater flow
Ocean Dumping	tidal motion (near shore)	earthquakes, wave motion

On-site treatment offers two major advantages over land disposal: 1) transport of hazardous substances is eliminated; and, 2) the problems of establishing and managing secure landfills are avoided.

The primary methods used for hazardous waste treatment are land application and incineration. Residues from both processes are often released into the environment by erosion or air currents. Harmful substances released by incineration are shown in Table 12-3.

12.8 Waste Reduction and Recycling

Concepts: source reduction, recycling

source reduction

Disposal and treatment are variations on the "end of the pipe" approach to pollution control. Truly comprehensive waste management also requires source reduction and recycling. Source reduction minimizes the production of the pollutant, while recycling reuses the material in its original form, or as a component in another process or product.

Source reduction offers significant potentials for reducing the risk from exposure to harmful substances. In order to realize these potentials, there are two major challenges: 1) providing adequate technology required to produce or identify non-toxic substances that can substitute effectively for current inputs; and, 2) there is often an "attitude adjustment" necessary on the part of the end-user (people) before source reduction efforts are successful.

For instance, white paper contains chlorinated hydrocarbons used in the bleaching (whitening) process. It is now possible to substitute recycled paper and other off-white inputs in the papermaking process, but this results in paper which isn't bright white. Until technology can find a non-toxic method for producing bright white paper, are people willing to accept the alternative?

TABLE 12-3 Pollutants Released by Incineration *

Pollutant	Description
Particulates	solid matter in soot or smoke
Sulfur dioxide	gas produced by burning organic materials such as leaves and vegetables
Nitrogen oxides	odorless gases produced during combustion of fossil fuels and organic materials
Hydrogen chloride	colorless, nonflammable gas produced when dyes, artificial silk, and paint are burned
Hydrogen fluoride	colorless gas or liquid with a strong odor
Carbon monoxide	colorless, odorless, and tasteless gas produced when materials containing carbon are burned
Arsenic	poisonous element found in insecticides, paints, and ceramics
Cadmium	toxic metal used in electroplating
Chromium	toxic metal used in chrome plating and copper stripping
Lead	toxic metal found in paints
Mercury	toxic metal used in industry and in agriculture as a fungicide
Dioxins and furans	highly toxic organic compounds; used in herbicides and organic synthesis

* for a mass-burn state-of-the-art garbage incinerator

Another benefit of source reduction is the effect is has on making treatment and disposal processes less harmful to ecosystems. For instance, paper is a major component of the solid waste stream, which when incinerated may release dioxins. If these compounds are never included in production, there is a reduction of the ecosystem impacts when they are burned.

recycling

Recycling also offers significant environmental benefits, particularly in terms of energy efficiency. As noted earlier (Chapter 7), energy efficiency is an indicator of the current balance between humans and ecosystems. Recycling reduces energy waste by reducing the energy required to mine new aluminum for cans, or mold new plastic for bottles. Table 12-4 shows the environmental benefits of recycling aluminum, steel, paper, and glass.

As with source reduction, recycling also requires public participation and attitude adjustments. Separation of wastes is a major factor contributing to the success of community-based recycling programs. This is why many cities facilitate separation by providing color-coded bins.

TABLE 12-4 Environmental Benefits of Recycling for Major Materials

Environmental Benefit	Aluminum	Steel	Paper	Glass
Reduction of:				
Energy Use	90-97	47-74	23-74	4-32
Air Pollution	95	85	74	20
Water Pollution	97	76	35	--
Mining Wastes	--	97	--	80
Water Use	--	40	58	50

Source: Robert C. Letcher and Mary T. Sheil. (1986). "Source Separation and Citizen Recycling," in William D. Robinson, ed., The Solid Waste Handbook, John Wiley and Sons, NY.

12.9 Summary: How Well Are We Doing?

Concept: integrated waste management

There is no single solution to the problems offered by hazardous wastes. At this point in time, technology cannot provide substitutions for every harmful substance, nor can the vulnerability of disposal sites to outside processes and the subsequent release of contamination be completely eliminated. There is a need to maximize the current potential for handling hazardous wastes by using multiple operations linked together in comprehensive programs.

integrated waste management

Integrated waste management represents this linkage of programs which consider the cost, loss, and impact from wastes. At the center of this approach are source reduction and recycling; these can be considered primary operations. Secondary operations include disposal and treatment. By removing contaminants from the waste stream, higher efficiencies of source reduction and recycling can significantly reduce the environmental side effects from the secondary operations—disposal and treatment.

Waste and the associated space used for waste management are components of ecosystems. Integrated waste management minimizes the ecosystem impacts by reducing the energy inputs for material acquisition and disposal, and by reducing the impacts to land, air, and water.

Sustainable development efforts should incorporate the concept of integrated waste management, or the energy better used for producing resources will instead be required to handle waste. This has special implications for LDCs, where energy efficiency can help make the transition to a consumer-based society more economical. In MDCs, the environmental benefits of integrated waste management would offer some relief for overburdened landfills and polluted groundwater.

Applications / Discussion

12.2) Using the life cycle framework of Table 12-1 shown below, what are the temporal and spatial characteristics of gasoline exposure?

Life Cycle Event	Frequency of Exposure	Areal Extent	Concentration within Affected Area
Production			
Distribution			
Use			
Secondary Dispersion			

12.3) Using a local land use map, soil map (or soil survey available from the Soil Conservation Service), and topographic map (e.g., a U.S. Geological Survey Quad), evaluate the siting of landfills near year city. Base your evaluations on the slope, soil suitability (drainage), and proximity to population and wetlands, including floodplains.

12.4) How can we improve our decision-making about the uncertain effects of hazardous substances? Hint: consider scientific research and communication.

12.5) There are four criteria used in the Resource Conservation and Recovery Act of

1976 for defining hazardous substances: ignitability, corrosivity, reactivity, and toxicity. If policy is adaptive, and reflects the compromise of many groups, how is it possible that biological considerations, such as whether or not the substance is a carcinogen, were omitted? How does the omission of biologic criteria affect hazardous waste monitoring programs?

12.6) If you had only enough available funds to site a hazardous waste disposal facility correctly, or to provide for comprehensive monitoring of the site, which one would you choose? Why?

12.7) Complete the table below:

Treatment	Geosphere Component	Potential Environmental Impacts
Land application of waste oil (low persistence)	biosphere	microflora damage
	lithosphere	
	hydrosphere	
	atmosphere	

12.8) Perform a simulation of source reduction and recycling on your household solid waste. Use the table below:

Guidelines:

a) under column 1, monitor your solid waste for 1 week and track the percentages by volume; at the end of the week, calculate the percentage for each waste category

b) under column 2, identify source reduction or recycling activities which could reduce the waste generated by each category. For instance, grass clippings are waste if they are put out for collection; they are not waste if used as mulch on the lawn. Try to identify activities which keep items out of the landfill.

c) under column 3, estimate the new percentage of volume for each category if the activities in column b were performed

How much reduction in waste was achieved overall? What category had the greatest percentage reduction?

	Current percent of volume	Source reduction/ recycling activities	Percent of volume after source reduction
Paper (newspaper and other)			
Cardboard (food packaging)			
Metal (cans)			
Organics (food, grass clippings)			

12.9) Is reduction of the population growth rate in LDCs sufficient to provide for sustainable waste management policies? Why or why not?

Recommended Reading

Bartimus, T., and S. McCartney. (1991). *Trinity's Children, Living Along America's Nuclear Highway.* Harcourt Brace Janovich, NY.

Carson, Rachel. (1962). *Silent Spring.* Houghton Mifflin, Boston, MA.

Rathje, W.L. (1989). "Rubbish!," *The Atlantic Monthly*, December.

Sunstein, C.R. (1990). *After the Rights Revolution: Reconceiving the Regulatory State.* Harvard University Press, Cambridge, MA.

Sources of Additional Information

Citizens Clearinghouse for Hazardous Wastes, 119 Rowell Ct., Falls Church, VA 22046, (703) 237-2249

Environmental Protection Agency, Community Planning and Right-to-Know Information Hotline (800) 535-0202 has information about the Toxics Release Inventory—a listing of hazardous sites.

INTERNET: Oak Ridge National Laboratory, Hazard Management Group
e-mail: jhs@ornl.gov

Labor Unions. The United Auto Workers and the Oil, Chemical, and Atomic Workers have health and safety information about workplace exposures to hazardous wastes.

References

Gots, R.E. (1993). *Toxic Risks: Science, Regulation, and Perception.* Lewis Publishers, Boca Raton, FL.

Hershkowitz, A. (1986). *Garbage Burning Lessons from Europe: Consensus and Controversy in Four European States.* Inform, NY.

Keller, E.A. (1992). *Environmental Geology* (6th ed.). MacMillan, NY.

Pollock, C. (1987). *Mining Urban Wastes: The Potential for Recycling*, Worldwatch Paper #76. Worldwatch Institute, Washington, D.C.

13

Soil, Land, and Land Use

Review

13.1 Summary

Soil, by nature of its formation and location, is at the crossroads of the geosphere. Residing at the top of the lithosphere, soil is formed largely by physical processes initiated within the hydrosphere, such as weathering. Soil also serves as a sink and pathway for moisture in the hydrologic cycle. In the biosphere, soil provides habitat for the most species, primarily insects. With respect to the atmosphere, soil color influences the amount of terrestrial reradiation, and soil temperature affects evaporation rates.

The processes underlying soil formation are dynamic; soil is a work in progress. The current status of soil formation represents a combination of the parent materials, e.g., shale, limestone, and granite, with the climate (temperature and moisture variations), vegetation, and slope of the area.

As the backbone of agriculture, soil must be managed for sustainable use. With population growth, more land—often of marginal agricultural value--is placed under cultivation. This results in land uses often grossly mismatched with the terrain, leading to severe erosion and sedimentation problems.

Worldwide, soil erosion is a major environmental and economic problem. If we are to sustain agriculture, and ultimately our population, greater attention needs to be placed on sustaining soil.

13.2 Geographic Organization of the Land

Concepts: land units, site

Consider this paradox: water imposes an order to land. How can that be?

Starting at the broadest scale—the entire earth--over seventy percent of the surface is water. In fact, earth might be a bad name for our planet; a more accurate name would be "water." But the name "earth" reflects the central role land plays supporting human life. We live on land. Earth was one large land mass before tectonic forces split and separated the pieces. After the breakup, continental land masses were the first level of organization imposed by water (the oceans and seas). It was now possible to identify the earth's land masses distinctly, e.g., Asia, Africa, etc.

Meanwhile, in the continental interiors, running water was being organized to do some necessary work. Precipitation falling on the continental masses had to go somewhere, and while the water traveled, it took part of the surface with it. Nature likes to economize, so running water took the easiest paths between the high and low ground as it flowed. The easiest paths were not always straight lines, because elevations and rock hardness varied along the way.

Water flows curved to accommodate these features of the landscape. After millions of years, the flowing surface waters carved out valleys and formed distinct pathways, or drainage networks within the areas they drained. Much of the land in these drainage areas, or watersheds, possesses distinct spatial and topographic characteristics.

Elevated land between the stream valleys is called upland; land in the lower elevations of the valleys is called lowland. Hills and slopes occur between these two main areas. These divisions represent another organization of land imposed by water.

land units

Within uplands and lowlands, some areas have similar water supplies, topography, soil, and other factors. These relatively homogeneous areas are called land units. Different types of human activity—land uses--are well suited to specific land units. For instance, large, level, and well-drained land units are good locations for mechanized commercial agriculture.

site

Land units can be thought of as specific locations, or sites, for human activity. The siting of most urban areas is related to water features (Table 13-1).

TABLE 13-1 Examples of Water Features and Urban Siting

Type	Examples
Confluence	Koblenz (Rhine-Moselle) ; Khartoum (White Nile-Blue Nile)
River gorge exit	Bonn (Rhine); Turnu Severin (Danube)
River bend	Volgograd (Volga); Magdeburg (Elbe)
Bridgepoint	London (Thames); Gloucester (Severn)
Fall line*	Richmond (Virginia); Augusta (Georgia)
Lake-river point	Chicago (Illinois); Rostov (Russia)
River promontory	Durham (England); Toledo (Spain)
Isthmus	Detroit (Michigan); Tunis (Tunisia)

* the fall line is a linear physical feature along which sharp drops in elevation occur, usually between a mountain range and coastal plain; e.g., the Appalachian Mountains and the Piedmont coastal plain of the eastern U.S.

adapted from Rugg, (1972).

Even at the small urban scale, (remember we started with the whole world), water imposes an order on the landscape.

13.3 Topography and Land Use

Concepts: topography, slope, topographic maps, mismatch of terrain and land use

topography

Topography refers to the configuration of the earth's surface and the position of its natural and artificial features. Slope is a key component of topography, and is expressed as the amount of elevation gain over a specified distance. A gain of 528 feet over a mile would represent a 10% slope (528 ft. / 5,280 ft. = 0.10).

topographic maps

Topographic maps are models used to depict the earth's surface features and slope. Shading is used to identify features, such as urban areas, and contour lines represent the elevation above sea level (Figure 13.1). The elevation specified between each contour line is the contour interval; allowing map users to calculate slope. Using the map scale, a user simply marks a straight line of a known distance and compares the endpoint elevations. With distance and elevation gain (or loss) known, the percent slope along the line can be calculated.

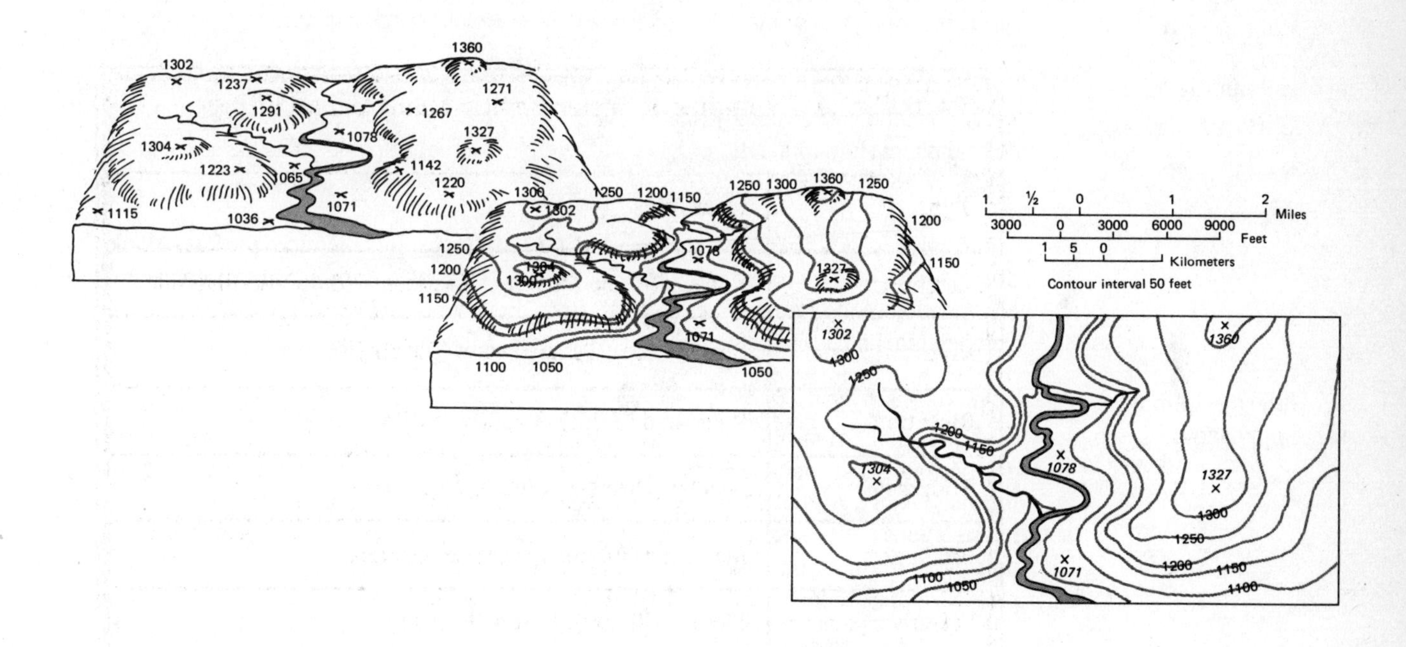

Figure 13.1 Topographic map

mismatch of terrain and land use

The mismatch between terrain and land use is often observed by analyzing topographic maps. In the map above, the ridges above the stream create an area of steep slope. This area is probably not suited to residential development because of the possibility of slope failure, e.g., high erosion and mudslides.

13.4 The Soil Mantle: Sources of Parent Material

Concepts: soil mantle, interdisciplinary approach

Soil can be thought of as a stage where physical, biological, and chemical processes perform. But not all soil came to the show in the same way. Some soil was chauffeured by wind and water, other soil didn't travel at all. In most cases, the soil formation process begins with the weathering of bedrock—the solid rock underlying all land. Weathering is the physical and chemical breakdown of rock, with water acting as the primary agent. As a physical agent, the force of running water splits rock and creates smaller fragments. Chemically, water dissolves minerals in rock such as calcium carbonate, which when added to water forms a slight acid (carbonic acid). The slightly acidic water dissolves more rock and carries it away. During this process, caverns may be created, resulting in karst topography. Eastern Europe and Mammoth Cave (Kentucky, USA) contain karst landscapes formed from this chemical process.

Some soil, called residual parent material, remains near the bedrock it was formed from. Other soil is taken by wind and water to new locations; these surface deposits become the raw materials for soil formation, and are termed transported parent material.

During all this upheaval, the soil formation process continues. Organic matter from plant decay mixes with the resident or transported parent material and forms a distinct upper layer called the solum. The solum supports life and overlies the subsoil, composed of the inorganic parent material which was either resident or transported.

soil mantle

The combination of the solum and the subsoil is called the soil mantle. The depth of the soil mantle varies roughly in accordance with slope: in stream valleys and low-lying coastal plains the soil mantle is deep, while in mountainous upland areas the soil mantle is thin.

The processes at work in soil formation are continuous. It is therefore useful to think of soil formation as work in progress. The current status of soil formation represents a combination of the parent materials, e.g., shale, limestone, and granite, with the climate (temperature and moisture variations), vegetation, and slope of the area.

interdisciplinary approach

Figure 13.2 shows the interaction of the soil-forming factors within three different environmental settings: a sunny slope, a valley bottom, and a shaded slope. As the figure demonstrates, soil characteristics will vary widely depending on the variations in the soil forming factors. All potential land uses should, at a minimum, consider the primary interrelationships of soil with the major environmental components. Whether building a pipeline to transport oil hundreds of miles, or to provide drainage for a single building, an interdisciplinary approach is required to study soil, use soil, and maintain soil.

13.5 Soil Properties: Key Soil Traits and Components

Concepts: texture, organic matter, ecosystems, infiltration, soil moisture balance

When you squeeze a handful of soil from a garden, field, or forest, it feels a certain way. Some soils feel grainy, some smooth, and others mucky. These sensations result from particles of three general sizes (sand, silt, clay) being present in the soil.

texture

Soil texture is the percentage of weight of each of these particles in a representative

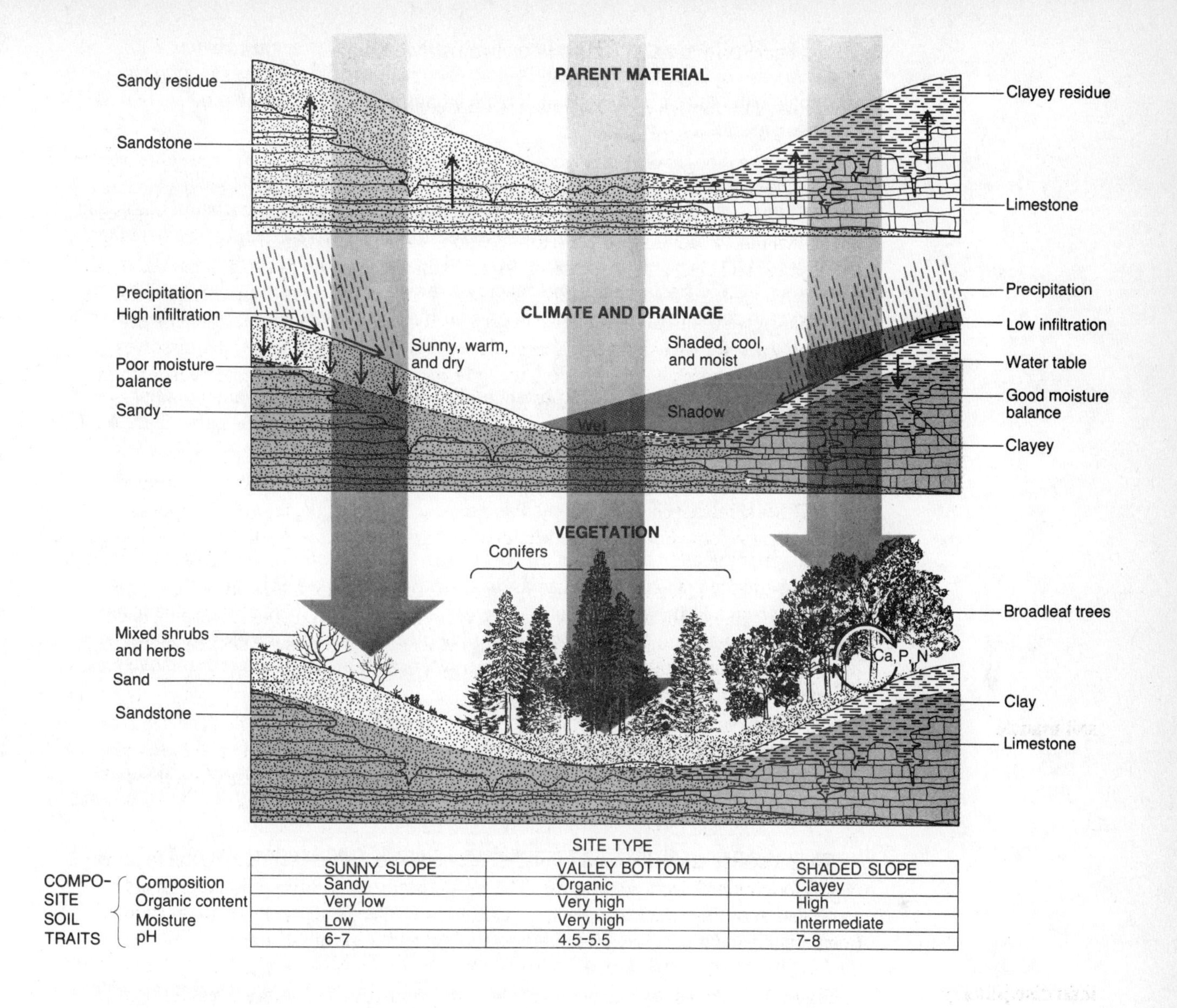

COMPOSITE SOIL TRAITS	SUNNY SLOPE	VALLEY BOTTOM	SHADED SLOPE
Composition	Sandy	Organic	Clayey
Organic content	Very low	Very high	High
Moisture	Low	Very high	Intermediate
pH	6-7	4.5-5.5	7-8

Figure 13.2 Soil forming factors in an environmental setting

sample of soil. In terms of particle size diameter, sand is the largest at 0.015 mm to 2 mm, silt is intermediate with sizes ranging from 0.002 mm to 0.05 mm, and clay is smallest with particles less than 0.002 mm.

Sand, silt, and clay have important roles in soil formation and influence other properties of soil, such as the movement and retention of water. For instance, sand improves soil drainage, while silt and clay improve a soil's water retention capability. Clay also helps soils to retain nutrients. The positive ionic charge of small clay particles attracts the negatively charged ions of the nutrients calcium and potassium, and the bonding between them prevents their removal and helps to maintain soil fertility.

organic matter

The level of soil fertility is related to the amount of organic matter present. Inputs of organic matter to soil come from plant cover, which is then decomposed and consumed by algae, fungi, and bacteria. This biological activity produces humus. Humus is a reduced form of organic matter, and is more chemically stable than the new organic material on the soil surface. Soils with more humus are generally more productive than lower-humus soils.

ecosystems

The production of humus demonstrates one of the many ecosystems within soil. Algae, fungi, and bacteria are eaten by insects and worms (primary consumers), which in turn are eaten by spiders and centipedes (secondary consumers). Other predators at the tertiary consumer level then eat the spiders and centipedes. Humus is produced by each consumer; but the fraction of organic energy returned to the soil diminishes rapidly from consumer to consumer. Most of the energy consumed is dissipated by the bodily heat and motion of the animals.

The amount of organic matter present in the soil also affects the available water capacity (AWC). Soils with higher organic matter content have significantly higher AWC than soils of similar texture with less organic matter. For example, the AWC of a silt loam soil containing 4% organic matter by weight is more than twice that of a silt loam containing 1% organic matter by weight.

The water retention capacity of organic matter has implications for diverse land uses, such as agriculture and urbanized areas. In agriculture, more water from each rainfall event can be made available to plants; whereas in urban areas, there are possible reductions of stormwater runoff. Sustainable farming practices include the plowing under of crop plant residues, and in urban areas homeowners are now advised not to remove grass clippings and raked leaves from their lawns.

infiltration

Besides retention, another important relationship between water and soil is the infiltration capacity. This is the rate of penetration expressed in millimeters or centimeters of surface water lost to the soil per hour. Infiltration rates vary with soil texture, plant cover, existing soil moisture, ground frost, and land use. Sand has high infiltration capacities, which explains why many athletic fields use sand as a base below their sodded areas.

soil moisture balance

Within the soil, water moves in three different paths. Capillary water is held around and between soil particles by cohesion among water molecules. This water is available to plants. Gravity water moves downward and eventually becomes groundwater; this water is fluid and not held by molecular forces. The third path is represented by the loss of water (usually capillary water), to evapotranspiration—a combination of evaporation from the soil and transpiration from plants. A soil is at field capacity when it can hold no more capillary water.

Different quantities of water move in and out of the soil throughout the year. Moisture flows between the atmosphere and soil, soil and plants, and plants and the atmosphere. Thus, precipitation is the only inflow of water to soil, and evapotranspiration the only outflow.

The soil moisture balance for any period of time is equal to precipitation minus evapotranspiration. Positive balances occur in soils within humid climates, and the excess water recharges groundwater. Soils in arid climates have negative moisture balances, and irrigation is often required to grow crops.

13.6 Soil Forming Processes: Solum as a Chemical System

Concept: soil profile

Water transports soil particles, minerals, and organic matter from different sources. This mix of materials forms a dynamic chemical system within soil, as new sources of matter are continuously provided by bedrock, vegetation, runoff, and the atmosphere.

Variations in soil moisture, temperature, composition, and the amount of ions serve as controls on the chemical processes occurring within soil. The movement of ions is particularly important to soil formation. As noted in the prior section, ions are atoms (e.g., potassium K+, Calcium, Ca++), or groups of atoms (e.g., carbonate, CO3-), which carry electrical charges. Water can move ions to different depths in the solum, with the controls previously mentioned determining the rate and amount of movement.

The removal of mineral ions from a higher zone to a lower zone within the soil is called leaching. If tiny clay particles, or colloids are also relocated, the term eluviation is used to describe the process. The zone of soil losing materials is called the zone of eluviation, and the receiving depth is called the zone of illuviation.

soil profile

Internal material transfers and chemical processes coupled with external environmental influences can create distinct layers, or horizons within the soil. The layers within horizons often have distinct colors, texture, and chemical composition. A sequence of horizons in a section of soil is called a soil profile, and is shown in Figure 13.3.

In general, four standard horizons are recognized from top to bottom. They are designated by the letters O, A, B, and C. The O horizon includes the topsoil and humus, where the organic content is often greater than 20 percent. The A horizon corresponds to the zone of eluviation, and the B horizon is the zone of illuviation. The C layer is transitional to the parent material; here the alteration by soil processes is less advanced.

Not all environments support the development of a four-layer horizon. Soil horizons may be absent in areas where there is a high rate of parent material deposition, as in floodplains and sand dunes. And, arid regions often exhibit only a calcic B layer.

13.7 Integrated Models of Soil, Land Use and Environment

Concepts: soil classification, the 7th Approximation

soil classification

Soil profile variability, and the geographic distribution of soils has generated many different soil classification schemes. The current scheme used by the United States Department of Agriculture is called the 7th Approximation.

the 7th Approximation

Under the 7th Approximation scheme, there are seven levels of classification. The broadest level is the order, which is subdivided into successively smaller geographical units of sub-orders, great groups, subgroups, families, series, and pedons. Figure 13.4 shows the ten orders of this scheme for the Western Hemisphere.

The soil distributions shown in the figure occur within the context of interacting natural and human conditions, called environmental regimes. Seven key regimes are: wet tropical, desert and grassland, midlatitude forest, permafrost tundra, wetland, floodplain, and cropland. Various land uses have been adapted to the features of each soil group

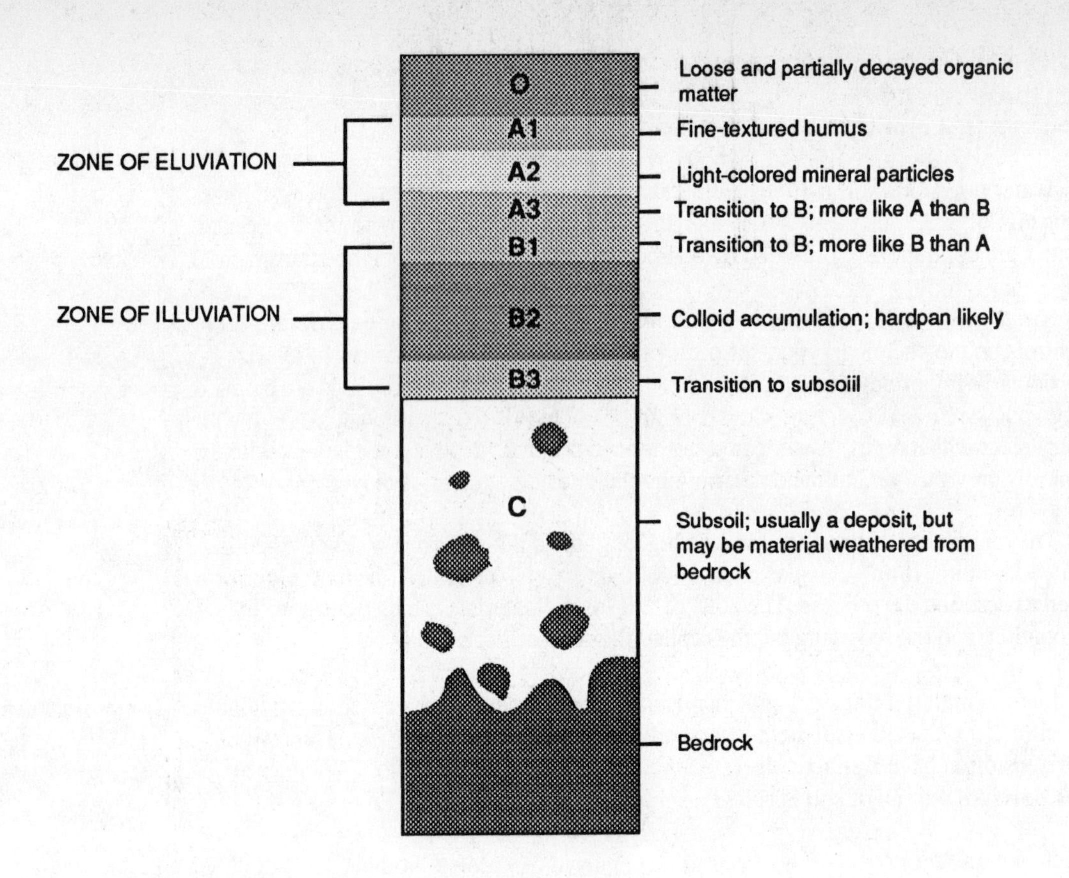

Figure 13.3 Soil profile

within environmental regimes (Table 13-2).

In the table, a "/" indicates there are significant differences in the features and/or land uses of particular soil groups. For instance, the desert/grassland regime has two major soil groups—aridsols and mollisols--which differ significantly in their abilities to hold moisture and support agriculture. In the midlatitude forest regime, the alfisols and spodosol groups have significantly different features, but support a similar land use.

13.8 Soil Loss By Erosion

Concepts: erosion, Universal Soil Loss Equation (USLE), erosion rate and sediment loss rate

erosion

Erosion is the dislodgement of soil particles by wind or water, with water causing the most soil loss. Five factors influence the rate of soil erosion: vegetative cover, rainfall energy, soil erodibility, slope, and crop management techniques.

Vegetative cover is the most important of the five factors contributing to the erosion rate. At the surface, vegetation stabilizes the soil and reduces erosion. Above the surface,

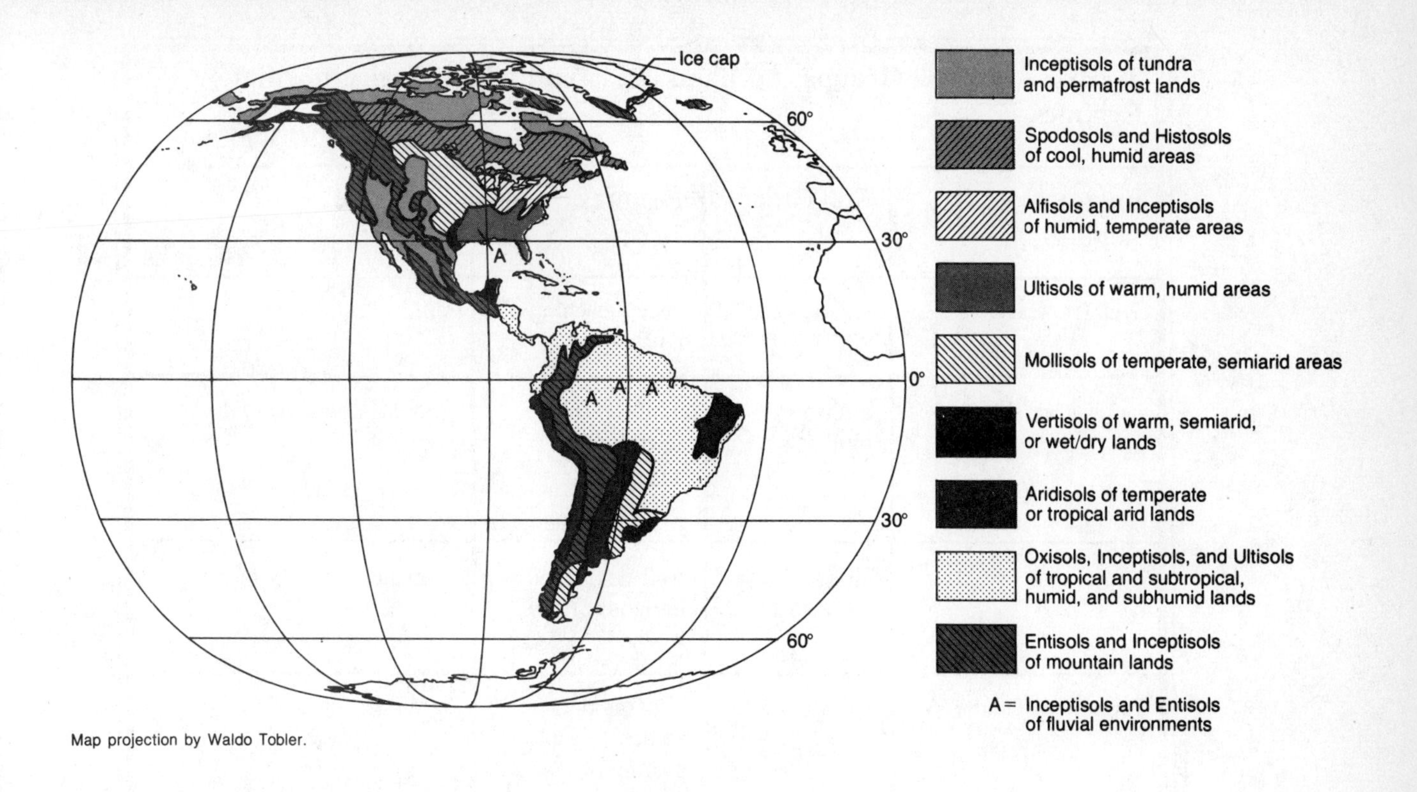

Figure 13.4 Soils of the Western Hemisphere, 7th Soil Approximation

vegetation foliage reduces wind velocities and intercepts rain, lowering their erosive capacities.

When land is disturbed by construction activity, soil erosion increases drastically. Erosion rates in urban areas as high as 40,000 times the preconstruction rate have been observed. The typical magnitude is between 20 and 100, which equates to a loss of 50 to 70 tons of soil per acre per year.

Soil erosion from agricultural lands is not as pronounced, with an average soil loss per acre per year of 2 to 4 tons. However, the failure to contour plow, the lack of a protective plant cover, sloping fields, and highly erodible loam or silt soils can produce much higher losses (200 to 400 tons per acre per year).

Universal Soil Loss Equation

The Universal Soil Loss Equation (USLE) is used to compute the potential soil loss from different land parcels by combining the five factors of erosion:

soil loss (tons/acre/year) = KRLSCP

where: K = soil erodibility factor
R = rainfall intensity
LS = slope length and steepness
C = vegetative cover
P = conservation/construction practices

Standard values of soil erodibility and rainfall intensity are readily available from the

TABLE 13-2 Soil Groups and Land Uses Within Key Environmental Regimes

Environmental Regime	Soil Group	Features	Land Use
Wet Tropical	oxisol, utisol	heavy leaching of nutrients	shifting cultivation
Desert/Grassland	aridisols / mollisols	light topsoils, salts / good moisture retention	grazing, nomadism / dryland farming
Midlatitude Forest	alfisols / spodosols	well-developed horizons / heavily leached	crop and livestock farming
Permafrost Tundra	inceptisols	high organic content, poor profile development	herding
Wetland	histosols	high organic content	often drained for agriculture
Floodplain	inceptisols	unstable, little horizon development, mix of clay, silt, and sand	agriculture, urban
Cropland	various	new textures from plowing, chemical inputs from pesticides and fertilizers, high erosion	crops

Soil Conservation Service; field observations are required for the other variables (Vig. 13.1).

Vignette 13.1 Quick Estimation of Erosion Potential

Sometimes you do not have the ability to obtain field data for erosion estimation, or the collection of these data for the entire study area is beyond the scope of the project. Instead, a conservative estimation technique can be used which employs values near the low end of the empirically determined erosion ranges for disturbed agricultural and urban lands.

Erosion estimation values

Activity/area	Erosion rate (tons/acre/year)
Urban construction	50.0
Agriculture	2.5
Stable urban areas	1.7

The goal of these estimates is not absolute accuracy; they serve as a quick method for estimating the relative proportions of agricultural to urban erosion potential and for obtaining a sense of the problem's magnitude in the study area. For the best estimation accuracy, a newer version of the USLE called the RUSLE (Revised Universal Soil Loss Equation) has been developed by the Soil Conservation Service (see Sources of Additional Information).

How could you increase the accuracy of the quick estimation method outlined above?

---

erosion rate and sediment loss rate

Most eroded soil is deposited in sinks within a watershed. Wetlands, swales, lakes, ponds, floodplains, and vegetated areas receive large quantities of eroded material. The actual erosion rate is therefore quite smaller than the amount of material exported outside of a watershed. This exported amount is called the sediment yield rate.

A typical ratio of these rates (erosion rate / sediment yield rate) in a watershed is 4.0; that is, there is 4 times more soil being eroded than being exported from the watershed. Management practices to control erosion should strive to lower the quantities of eroded soil deposited within and exported out of the watershed.

13.9 Environmental Impacts of Soil Erosion

Concepts: land degradation, environmental costs

land degradation

There are several primary impacts of erosion worldwide: crop loss, declining agricultural fertility, sedimentation, and wetland eradication. The agricultural impacts from erosion are primarily the result of topsoil removal, while the other forms of land degradation reflect the deposition of eroded material.

In all nations, soil erosion causes extensive environmental damage. This is especially apparent in the LDCs, where soil erosion is often the key component of a general pattern of land degradation (Table 13-3).

TABLE 13-3 Summary of National Environmental Damages from Land Degradation in Selected Developing Countries

Country	Environmental Damage	Year	% GNP *
Burkina Faso	Crop, livestock, and fuelwood losses	1988	8.8
Costa Rica	Coastal fisheries destruction soil erosion, deforestation	1989	7.7
Ethiopia	Effects of deforestation on fuelwood supply and crop output	1983	6.0-9.0
Indonesia	Soil erosion and deforestation	1984	4.0
Madagascar	Land burning and soil erosion	1988	5.0-15.0
Malawi	Soil erosion and deforestation	1988	2.8-15.2
Mali	Effects of soil erosion on crop output	1988	0.9-12.5
Nigeria	Soil degradation, deforestation, water pollution, other erosion	1989	17.4

* GNP = Gross National Product

Source: Barbier, E.B., and J.T. Bishop (1995).

environmental costs

As the table demonstrates, the environmental and economic costs of soil erosion are staggering in some nations. Almost one-fifth of Nigeria's total value of goods and services is lost from land degradation, of which soil erosion is a major component. This demonstrates why a sustainable economy must include sustainable land uses.

13.10 Summary: How Well Are We Doing?

Concept: spatial and temporal aspects of land management

The last section demonstrated the severe problems facing some LDCs with respect to soil erosion. The erosion rate in MDCs is about one third the LDC rate, but this is still a substantial amount.

spatial and temporal aspects of land management

Land and soil erosion management can be evaluated within the context of spatial and temporal factors. Spatially, there is an increasing expansion into marginal lands, which are less productive, often more prone to floods and drought, and lack productive soil.

Moreover, development costs per acre are higher in those lands which are less accessible and in a poorer state initially.

The temporal factors of human management reflect increasing frequencies of human intervention. Most of these activities have negative impacts, such as the more frequent fertilization of over-farmed land designed to offset its declining productivity. There is also the use of shorter fallow periods, which prevents the full regeneration of organic nutrients into the soil.

Population growth has contributed to these trends, as the need to feed more people has generated the expansion into marginal lands and a more intensive use of existing land. In order to ensure the sustainability of agricultural systems, and ultimately the future of our population, soil must be managed for sustainability. We are a long way from this goal.

Applications / Discussion

13.2 - 13.3) Using a topographic map (a USGS 7.5 minute quad will do), identify the major land units within the map area. Use water supplies, topography, soil, and any other suitable criteria.

13.4) Is the soil in your area derived from residual or transported parent materials? Using a geological map showing your area's bedrock, and considering the general moisture and temperature variations, what properties would you expect the soil to have, e.g., texture and mineral composition?

13.5) Using a garden hose, water a small section of a lawn until water pools on the surface. Observe the infiltration into the soil. What factors are affecting the rate of infiltration at this site?

13.6 - 13.7) Obtain a soil survey of your area; in the United States these are performed by the Soil Conservation Service. Perform these steps:

1) identify the major soils in a subsection of a county; an area of 2-4 square miles is sufficient

2) record the major characteristics of these soils—erodibility, texture, infiltration capacity. The narrative at the beginning of the survey book should contain this information.

3) using land use and topographic maps, compare the land uses and slopes with the characteristics of the soils existing within each major land use—agriculture, urban, commercial, recreation, or industrial

Are there any mismatches between a particular soil's suitability to sustain a given land use, and the existing land use?

13.8) What would a low soil erosion to sediment yield ratio indicate about the physical features within the watershed? What would be the climatic characteristics of a watershed where this ratio was close to 1.0?

13.9) What are the primary environmental impacts of erosion in your area? Hint: consider the four spheres of the geosphere.

13.10) How are land and water management related? If land management is improved, what implications does this have for water quality? Cite specific examples.

Recommended Reading

Brown, L.R. and E.C. Wolf. (1984). *Soil Erosion: Quiet Crisis in the World Economy*, Worldwatch Paper #60. Worldwatch Institute. Washington, D.C.

Pearce, D.W., E.B. Barbier, and A. Markandya. (1990). *Sustainable Development: Economics and Environment in the Third World.* Earthscan, London.

Sources of Additional Information

INTERNET: Soil Conservation Service Gophers / Soil Conservation News on the North American menu

Lal, R., (ed.). (1988). Soil Erosion Research Methods. Soil and Water Conservation Society, Ankeny, IA.

Revised Universal Soil Loss Equation (RUSLE). A computer program available from the Soil and Water Conservation Society, (800) THE-SOIL

The World Bank, Washington, D.C.

References

Barbier, E. B., and J. T. Bishop. (1995). "Economic Values and Incentives Affecting Soil and Water Conservation in Developing Countries," *Journal of Soil and Water Conservation* (50), 133-137.

Buckman, H.O., and N.C. Brady. (1974). *The Nature and Property of Soils* (8th ed.). Macmillan, New York, NY.

Goldman, S.J., K. Jackson, and T.A. Bursztynsky. (1986). *Erosion and Sediment Control Handbook.* McGraw-Hill, New York, NY.

Hudson, Berman, D. (1994). "Soil Organic Matter and Available Water Capacity," *Journal of Soil and Water Conservation* (49), 189-194.

Marsh, W.M., and J. Dozier. (1981). *Landscape: An Introduction to Physical Geography.* Wiley, NY.

Rugg, D.S. (1972). *Spatial Foundations of Urbanism.* William C. Brown., Dubuque IA.

Wolman, M.G., and A.P. Schick. (1967). "Effects of Construction on Fluvial Sediment, Urban and Suburban Areas of Maryland," *Water Resources Research* (3), 451-464.

14

BIOLOGICAL DIVERSITY AND LAND USE

Review

14.1 Summary

Earth has been a busy place the last 3 million years. Our climate changed several times, and the planet got a face lift when the continents split apart. In addition, company arrived—new life forms emerged from the sea and took up residence on the land.

Over millions of years, the millions of distinct organisms on earth became organized into species. Species are populations reproductively isolated from all others. Each species also developed its own needs for space, food, and shelter. Some species shared the same space, but interacted with their environment, or habitat, differently.

Occasionally, large-scale physical disturbances disrupted habitats and caused the extinction of many species. Six of these disturbances have been documented in the last 550 million years. These extinction episodes have caused short-term geological fluctuations in biological diversity—the total number of species inhabiting the earth at a given time.

When human landscape modification began 11,000 years ago, there was increased pressure on species habitat and species diversity at different scales. Over large areas, agriculture replaced native plants, new species were introduced into new habitats, and in recent years, increased rates of tropical rainforest destruction has threatened or eliminated species with ranges limited to several meters.

Biodiversity on earth has offered, and will continue to offer, new sources of food and medicine. But the functional diversity inherent to a species-rich landscape also sustains our basic life support systems—ecosystems. In order to maintain biodiversity, proper land management and cooperation of many institutions across different political boundaries are required.

14.2 The Concept of a Species

Concepts: species, speciation, range, extinction

species

Species are populations reproductively isolated from all others. Reproductive isolation means distinct species members will breed among themselves, but not interbreed with members of other species.

speciation

There are two basic theories of species formation, or speciation. One is based on geographic processes, the other is related to population social structure and chromosonal change. The geographic view is more traditional, and it states that new species form when an ancestral population is split by a geographic barrier, such as a mountain range or river gorge. These two descendant populations then evolve independently and form a new species. The evolution of the two new species occurs by adaptation to the different environments of their separated places. After a long time, the populations become so different they are incapable of interbreeding should contact become reestablished.

The social structure and chromosonal change theory of speciation attributes the origin of some new species to accidental chromosonal change. This can occur where populations are subdivided into small groups of kin that breed exclusively with each other for several generations. Harems have these attributes, and they consist of a dominant male maintaining several females, and brother-sister mating among the offspring. If a chromosonal mutation arises in the dominant male, this can generate a reproductive chain reaction leading to the formation of a new species. This process is depicted in Figure 14.1.

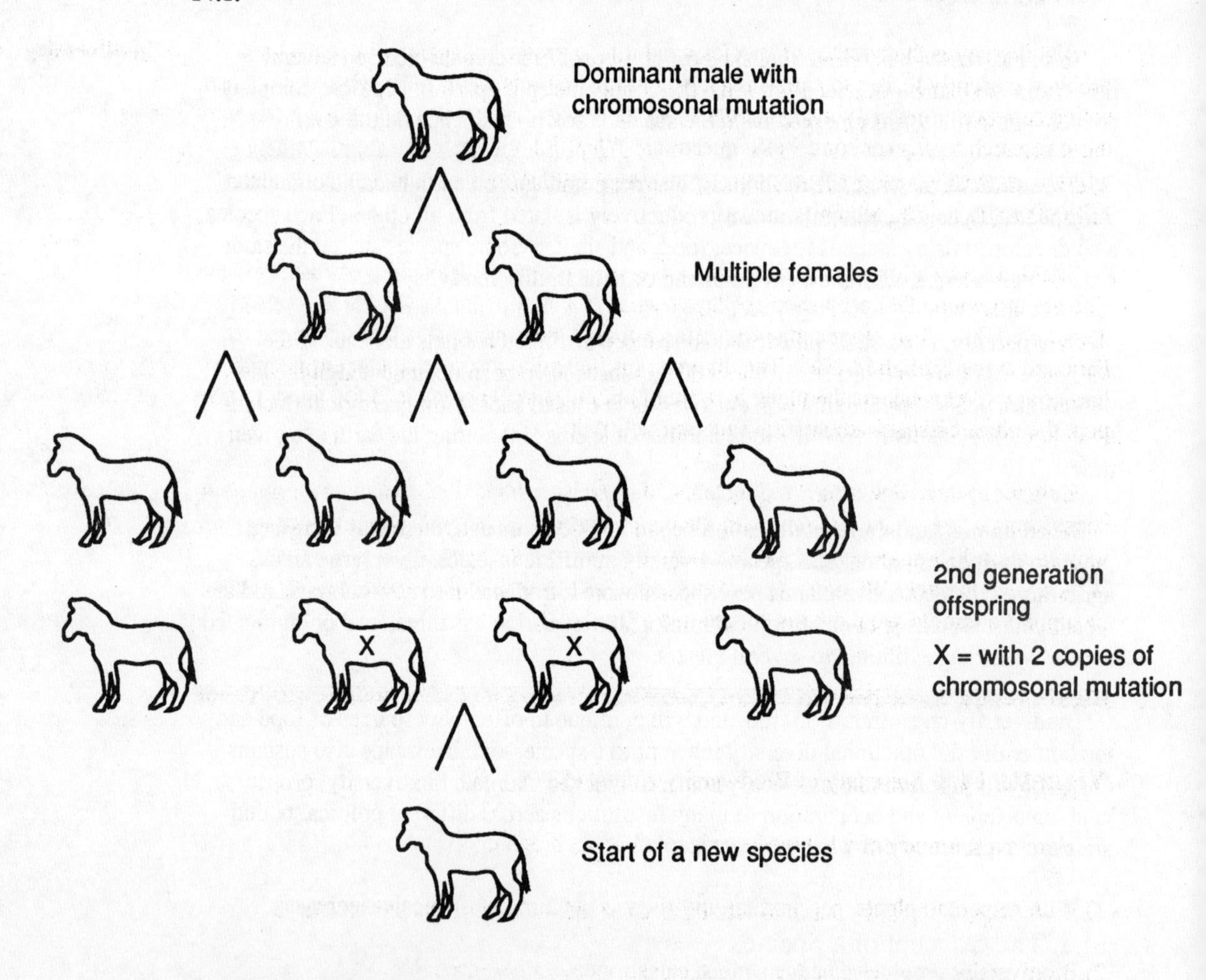

Figure 14.1 Chromosonal speciation

Horses provide a strong circumstantial case for chromosonal speciation. All horses maintain a harem structure for kin breeding. The seven living species of horses (two horses, two asses, three zebras) have strong physical and behavioral similarities—the differences are mainly superficial (color and pattern). But the number of chromosomes differ greatly within these species; for instance, one zebra has thirty-two, while one horse has sixty-six.

range

Besides genetic differentiation, another important characteristic of species is their range. The range of a species is defined by the geographic area it occupies. Species range may be world-wide (cosmopolitan), or limited to a very small geographic extent (endemic).

extinction

The range of a species has an important relationship to its vulnerability to extinction. Some species are restricted to tree canopies or islands, and human land clearing activity within these areas can remove the total local environment and cause extinction.

14.3 The Significance of Biodiversity

Concept: biodiversity

biodiversity

Biodiversity is the number of species inhabiting a prescribed geographic area. One problem currently facing scientists regarding biodiversity is the need to produce evidence which supports previously made claims about its benefits. There is a requirement for more research to answer some basic questions: What if a few species become extinct—will the ecosystems adapt? Are some species more important than others in ecosystems? If biodiversity is lost, what else is lost?

The debate over the significance of biodiversity reflects two general positions. On one side are those who feel each species plays a small but significant role in the working of the whole. This view, often called the "rivet-popper" hypothesis, was first advanced by Paul and Anne Ehrlich in 1981. The diversity of life is like the rivets on an airplane; the loss of any rivet weakens the plane by a small but measurable amount. If too many rivets pop, the plane crashes—some vital function will fail.

Counter to this view is the "redundancy" hypotheses, which believes not everything in ecosystems is important. Staying with the plane analogy, most species in this scenario are viewed more like passengers than rivets—it is a few drivers keeping the plane flying. Which view is right? We don't know for certain, as the results from recent research seem to support a middle ground between the two (Vig. 14.1).

Vignette 14.1 Summary of Biodiversity Knowledge

Here's a summary of what we have learned about biodiversity:

1) With respect to plants, net productivity rises as the number of species increases.

2) Biodiversity provides a hedge against catastrophe, such as drought.

3) Species diversity carries with it functional diversity. A change in biodiversity can affect various ecosystem processes differently, such as water retention and nutrient cycling.

4) Ecosystems may be able to tolerate some random loss of biodiversity, but human-induced losses are far from random and are more likely to alter ecosystem processes.

5) Scale affects biodiversity and ecosystem processes. Some species have localized effects. Trees fall and create sunny gaps for other vegetation; earthworm casts create hot spots for denitrification in the soil. Across a landscape, the complex interactions of many species tends to average out to a functional constancy that can sustain some loss of individual species.

6) Not every possible functional niche is filled in every natural ecosystem. Some bromeliads fix CO_2 at night in the tropical forests of South America, but in the forests of Africa and Asia, nothing draws carbon from the atmosphere during the night. (Bromeliads are epiphytic plants—they obtain their nutrients and moisture from the atmosphere and rain directly, and usually grow on top of other plants, e.g., Spanish moss).

7) Ecosystems are in a state of constant change, and can learn to work with new species. The key question is whether humans can live within the new systems.

8) Many species have made contributions to medicine and agriculture. Over forty percent of prescription drugs come from natural sources, and grains known to indigenous peoples, such as amaranth, have excellent nutritional value. What we fail to learn about the thousands of unclassified plant species can only hurt us.

9) We don't know much. One of the largest gaps in our ecosystem knowledge involves the microbial world. Between seventy and eighty percent of the organisms in soil are unknown; this figure is over ninety percent for the oceans. In the environments of deep ocean trenches, hot geysers, and arctic deserts, microorganisms are the main component of ecosystems.

The issue of whether everything in ecosystems is important may not be answered for decades. Yet this question, and many others must be answered. If biodiversity is eliminated, we will never know the processes at work within them, or their potential benefits.

How much should society invest in the preservation of biodiversity--as much as the space program? Military budget? Cancer research? Genetic engineering?

14.4 Species Counts, Estimates, and Extinction Trends

Concepts: species counts, extinction

species counts

Current estimates place the number of species on earth at 1.4 million. Most are insects (750,000), but many more within this group, and additional species of plants, microorganisms, and invertebrates exist. Areas with uncharted biodiversity—the oceans and soils, contain thousands of unrecorded species.

extinction

Over geologic time, there have been short-term fluctuations in biodiversity. Six major extinction events have been identified: 1) at the end of the Pre-Cambrian (~515 Ma, Ma = million years ago); 2) at the end of the Ordovician (~438 Ma); 3) the late Devonian (~367

Ma); 4) at the end of the Permian period (~245 Ma)—generally recognized as the most severe; 5) at the end of the Triassic (~208 Ma); and, 6) at the end of the Cretaceous period (~65 Ma).

The causes of mass extinction are unknown. The only common factor is a declining sea level. Falling seas would have major impacts on biodiversity by draining the species-rich continental shelves. However, the rapid rate (in geologic time) and extent of species decline has led scientists to look for external causes, such as asteroid impacts.

Extinction is a normal consequence of not being able to adapt to new environments, but the wide influence of humans on ecosystems has also contributed to species extinction. The onset of agriculture eradicated many species of plants, while the destruction of tropical rainforests has threatened, and perhaps eradicated an unknown number of species.

<u>What role do zoos play in species preservation? Would it be better to try and preserve all species in their natural habitats? What factors affect this decision</u>?

14.5 Geographical Biodiversity

Concept: island biogeography

Biogeography studies the geographical distributions of species; island biogeography is a sub-field of biogeography concerned with the interrelationships between species distributions, island area, distance between land masses, and biodiversity.

island biogeography

Area and distance have significant effects on island biodiversity. Larger island size means the island can support more habitats and consequently more species. More numerous habitats may also promote species persistence by acting as refugia during environmental disturbances. In terms of distance, greater distances between islands, or between islands and the mainland are associated with lower rates of biodiversity, due to the reduced opportunities for genetic mixing.

The principles of island biogeography may lend some insights into the effects of human impacts on other terrestrial ecosystems. Forest clearing often results in patchy areas of vegetation and greater distances between their ecosystems, creating some spatial arrangements similar to those found between islands and the mainland.

14.6 Land Use, Habitat Loss, and Biodiversity

Concepts: habitat, scientific method, deduction, induction

Habitat consists of the environmental features which support organisms. Soil, micro-climate, cover, available food supply, and the interaction with other organisms are the principal factors determining habitat. All organisms interact differently with their own habitats, although they may share the same space.

habitat

At this point, we can conduct a closer examination of the similarities between biodiversity of islands and patchy areas on the mainland. Let's be blunt—habitat islands on land and true islands are not completely alike, which makes any conclusion from their comparison susceptible to error. Why?

Before we apply a general theory to explain specific events in new locations, there should be: 1) a thorough investigation of the processes operating in each environment; 2) a high level of congruence between the processes existing within the original environment and the new case—here, congruence means similar cause and effect relationships, and; 3) an accurate measurement of these processes within both environments.

scientific method

For instance, it would be unscientific to use the theory of "survival of the fittest" in nature to explain why certain people in society perform better than others. Many of the internal and external processes of nature and society are completely different. For example, the longevity of particular species in nature often depends directly on predator-prey relationships, but this is not a fundamental process affecting the human lifespan. And, the predictors used for social attainment, such as intelligence, are often inaccurately measured (if they can be measured at all).

With respect to habitats, patchy land and island habitats are not strictly comparable. To demonstrate why, let's begin with a basic component of the scientific method—observation. Figure 14.2 juxtaposes an island group with patchy areas of a tropical rainforest brought about by human deforestation.

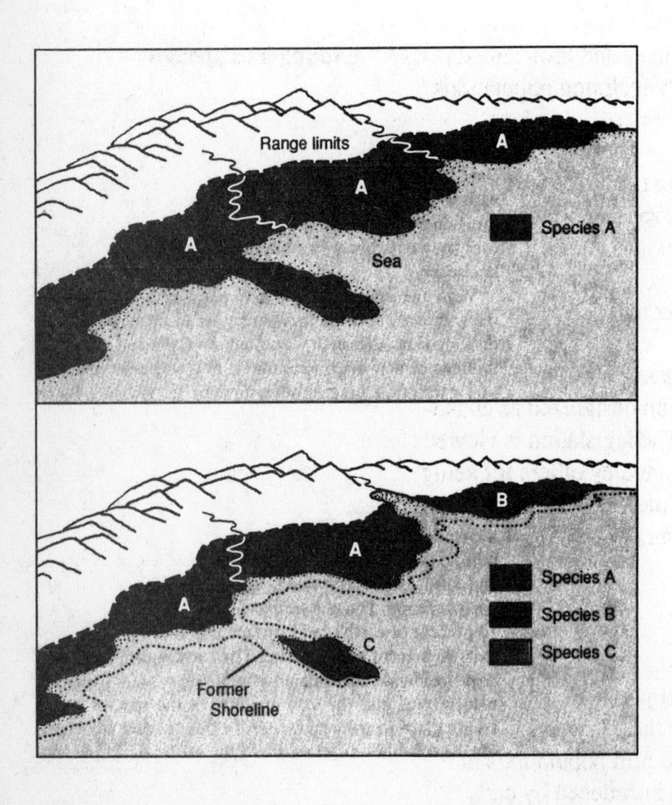

Figure 14.2 Habitat comparisons: islands and habitat islands on land

In the figure, spatial differences between the two habitats are apparent, and this would affect the environmental processes on both. One example of this is found in their edge effects—those processes occurring at the boundaries of the two locations. Islands are surrounded by aquatic ecosystems, while patchy land is bordered by terrestrial ecosystems. These physical differences would create basic differences in food supply and microclimate.

Considering other processes, organism mobility on islands and patchy landscapes also differs. In nature, different species have different ranges; e.g., some are migratory, others nonmigratory. The potential to move and establish new habitats after disturbance is greater for continental-based terrestrial organisms than island-based terrestrial organisms.

These examples illustrate the care which must be used when applying established theories to new topical or geographical areas.

deduction, induction

Application of general theory to explain the behavior of specific cases is called deduction, as opposed to induction, which is the process of forming inferences about a whole class from observations on a few of its members. Merely knowing these definitions is not enough to practice good science—the ground rules for their application must also be followed.

14.7 Endangered, Threatened, and Protected Species

Concepts: endangered species, exotic species

endangered species

Many MDCs and LDCs are trying to reduce species loss. Policies and laws aimed at protecting species near extinction (endangered), and species with declining populations (threatened), have been enacted.

At the international level, several organizations are working to prevent extinctions. One of the most effective is the International Union for the Conservation of Nature and Natural Resources (IUCN). The IUCN encourages nations to conduct inventories of plants and animals within their borders, and advocates the training of plant and animal biologists along with the establishment of nature preserves to protect endangered species.

In the U.S., protection of endangered species and threatened species is regulated by the Endangered Species Act of 1973. About 300 species have been designated as endangered, although not everybody is happy with this arrangement. The legislation is viewed by some as too broad and a hindrance to economic development, and by others for being too narrow because it focuses on individual species instead of protecting biodiversity and habitats. The act essentially reflects our uncertainty about biodiversity, and as more is learned, the regulations will improve.

Here is what we do know about extinction:

exotic species

The vulnerability to extinction is highest in species with small populations, or those species in isolated environments. Whales breed in groups, and if the group becomes too small, breeding may cease. Island species, including many native bird populations in New Zealand and Hawaii, were decimated by predatory animals introduced by early settlers. Non-native species, or exotic species pose significant threats to entire ecosystems. One reason for this is the lack of natural population control mechanisms in the new ecosystem—predators may be absent.

In the Great Lakes since 1800, at least 139 nonindigenous fishes, invertebrates, algae, and plants have been transported into the ecosystem by humans (Fig. 14.3).

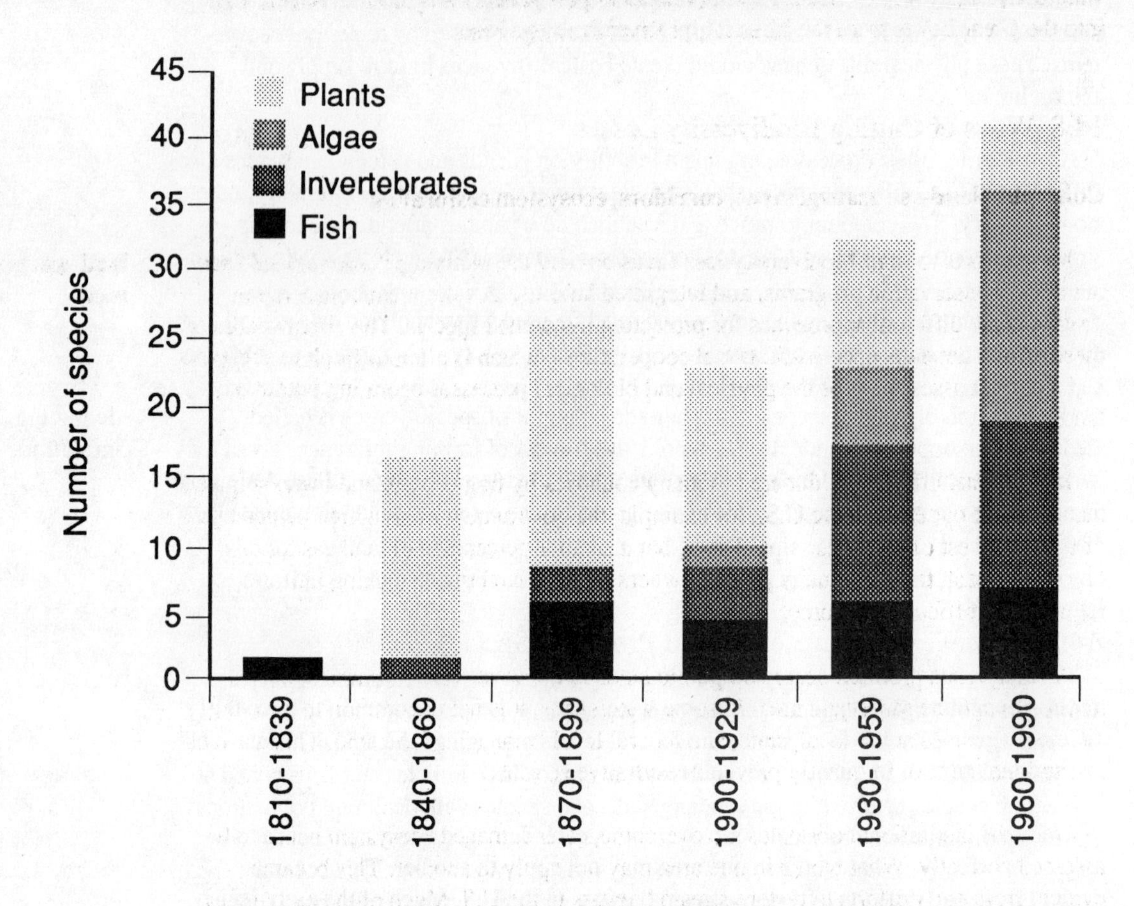

Figure 14.3 Timeline of Great Lakes introductions of non-native species

A variety of mechanisms have existed for non-native species entry into the Great Lakes. Canals built in the early 1800's linked drainage basins with the Great Lakes, and provided a conduit for organism migration. Ships entering the Great Lakes in the early 1800's used solid ballast (stabilization), including rocks, mud, or sand. This enabled many plants and invertebrates to be transported large distances to North America. Often attached to the hulls of these ships were sea lampreys, a particularly destructive non-indigenous fish which continues to prey on commercial fish in the Great Lakes.

After the 1930's, more ships began using water for ballast. With the opening of the St. Lawrence Seaway in 1959, the increase in transoceanic trade increased the number of ships entering the Great Lakes and the amount of ballast water released. At this time the zebra mussel, currently the most troublesome invader, was introduced. Zebra mussels multiply rapidly, and block the intake pipes of municipal water supplies. In Lake Erie, densities of 27,000 zebra mussels per square meter (their average size is equivalent to the tip of your small finger) have been documented.

Other conduits for non-native species in the Great Lakes have been the railroads, government agencies, and the aquarium trade. Railroad building disturbed habitats and promoted the dispersal of plants. Government agencies have deliberately introduced fish species in order to enhance fisheries, but associated organisms, many of them pathogens, also came along for the ride. Aquarium owners often discard their fish into the nearest waterbody. The mystery banded snail (*Viviparus georgianus*) was introduced this way into the Great Lakes from the Mississippi River drainage basin.

14.8 Ways of Cutting Biodiversity Losses

Concepts: land-use management, corridors, ecosystem restoration

land-use management

Most efforts to limit biodiversity loss focus on land use planning. National land use planning, conservation programs, and integrated land-use development/conservation programs are different approaches for protecting threatened species. The effectiveness of these efforts depends upon institutional cooperation—which is often difficult to achieve--and a correct assessment of the physical and biological processes operating within the landscape.

Institutional efforts are hindered in many countries by fragmented land ownership and management patterns. In the U.S., for example, the government owns a high percentage of the land west of the Mississippi River, but a smaller percentage of land east of this line. In the east, there are many private owners of large land tracts, making uniform regulations difficult to enforce.

An analogous problem occurs on public lands in the west, where several agencies might manage the same land area. In some watersheds, it is not uncommon to have thirty different agencies at the local, state, and federal level "managing" the land. This maze of institutional control frequently prevents restorative actions.

Once the institutional obstacles are overcome, each damaged ecosystem needs to be assessed correctly. What works in one area may not apply to another. This became evident from early efforts to restore stream habitats in the U.S. Much of the early focus was placed on restoring trout habitat, but since these fish are only part of the total stream ecosystem, restorative efforts had limited success.

corridors

More comprehensive approaches to habitat restoration are now emphasizing the linkage of land tracts by natural organism pathways, or corridors. Linkage of these areas can help overcome fragmentation and provide larger habitat areas capable of supporting more species.

ecosystem restoration

One example of this new approach is the Rapid Biotic and Ecosystem Response Strategy (RBER). This method advocates a combination of geomorphology, ecosystem science, and conservation biology to restore damaged watershed ecosystems. This broad approach to restoration begins with the identification and stabilization of areas with limited habitat damage called "hot spots," and proceeds by using corridors to link these areas with adjacent areas capable of becoming hot spots themselves. Restorative actions include comprehensive control of erosion and pollution from urban and agricultural sources, the use of riparian buffers, and the prevention of channelization, diking, and other artificial in-stream modifications.

Inertia from the entrenched engineering practices which have relied solely on structural in-stream modifications will continue to pose a considerable roadblock to new stream restoration approaches. However, as the number of undamaged stream miles in many countries dwindles and species continue to be threatened, the prospects for interdisciplinary habitat restoration efforts in watersheds and other areas will improve.

14.9 Summary: How Well Are We Doing?

Concept: environmental perception

The threat to biodiversity comes primarily from habitat loss and the introduction of exotic species. Economic exploitation cannot be ignored as a cause, but it is relatively minor compared with the other two impacts.

Technological change underlies much of the newer threats to biodiversity. Improved modes of transportation have increased the access to remote areas, and the land cleared for roads has fragmented large tropical rainforest regions in many nations.

environmental perception

Technology is also creating a barrier between people and the natural environment. People in MDCs now drive to most destinations, and environmental perceptions are now formed through the windshield of a moving car, not by direct exposure. Telecommuting is also limiting exposure to different environments, since millions are now working at home.

Reducing our direct environmental contact can alter our perceptions of the environment. We should guard against the "we can go anywhere, and do anything" mentality with respect to ecosystems. Armed with only an incomplete knowledge of species diversity and the human impacts on that diversity, we can achieve increasing contacts with ecosystems through scientific research and sustainable land use practices.

Applications / Discussion

14.2) Under what circumstances could the possibility of speciation by chromosonal change be enhanced by a geographic barrier?

14.3) What types of information about a specific ecosystem would be required to verify the "rivet-popper" hypothesis?

14.4) What are the major features of the asteroid theory of extinction? For reference, use the article by L.W. Alvarez, and F. Alvarez et al: "Extraterrestrial Cause for the Cretaceous-Tertiary Extinction," *Science*, vol. 208, 1980, pp. 1095-1108. Can you find any weaknesses in their theory?

14.5) Besides the increase in habitat refugia, what are some other advantages offered to wildlife by larger islands?

14.6) In your neighborhood you obtain a sample of thirty red ants of the same species, which also exist in several other states or provinces. After careful measurement of each ant, you calculate the average size as 8.00 mm. You repeat the test in the same location 10 times and obtain the same average size. Can you now make a strong inference about the size of this red ant species? If not, design a sampling procedure that would allow you to make a better inference about the size of this red ant species.

14.7) Under what circumstances would it be possible for an organism to exist as an exotic species only?

14.8) Why does habitat restoration require an interdisciplinary approach? List the scientific specialties required for restoration of a shoreline area damaged by an offshore (oceanic) oil spill. It may help to refer to newspaper articles about the oil spill in the Prince William Sound off the coast of Alaska which occurred on March 24, 1989.

14.9) Using the table below, make an inventory of your direct environmental contact for one week. Record the time with a precision to the 0.25 hour. For instance, if you spent 95 minutes in a car on Monday, record 1.5 hours in the appropriate cell.

	Time spent in a car	Time spent indoors	Time outdoors within 2 miles of home	Time outdoors > 2 miles from home
Monday				
Tuesday				
Wednesday				
Thursday				
Friday				
Saturday				
Sunday				

What percentage of your time is spent in direct contact with the natural environment? Do you think this percentage will increase or decrease as you get older? Why?

Recommended Reading

Darwin, C. (1859). *On the Origin of Species by Means of Natural Selection.* John Murray, London.

Ehrlich. P.R. (1986). *The Machinery of Nature.* Simon and Schuster, NY.

Sources of Additional Information

International Joint Commission on the Great Lakes, Windsor, Ontario.

INTERNET: Biodiversity Journals and Newsletters Gopher on the North American menu

Wilderness Society, 900 17th St. NW, Washington, D.C. 20006 (publishes a newsletter called "Green Fire" providing current information about the Endangered Species Act).

References

Baskin, Y. (1994). "Ecosystem Function of Biodiversity," *Bioscience* (44), 657-660.

Doppelt, B. (1993). *Entering the Watershed: a New Approach to Save America's River Ecosystems.* Island Press, Washington, D.C.

Gould, Stephen Jay. (1984). *Hen's Teeth and Horses's Toes: Further Reflections in Natural History.* Norton, New York, NY.

MacArthur, R.H., and E.O. Wilson. (1967). *The Theory of Island Biogeography.* Princeton Univ. Press, Princeton, NJ.

Mills, E.L., J.H. Leach, et al. (1994). "Exotic Species and the Integrity of the Great Lakes," *Bioscience* (44), 666-676.

Signor, P.W. (1994). "Biodiversity in Geological Time," *American Zoologist* (34), 23-32.

15

OPEN LAND RESOURCES: FORESTS, RANGELANDS, PARKS, AND PRESERVES

Review

15.1 Summary

Open lands are ecosystems consisting of forests, rangelands, woodlands, and tundra. Some areas have been exploited at great human and environmental cost. Others have been preserved, but not without great effort.

The debate over open space management centers around preservation and use. Do we set aside these areas, or allow them to be developed? What approaches have been applied, and is a balance between them possible?

Open spaces of forest and range have natural and economic values. Sustainability of these values depends upon the ability of society to balance its conceived values of preservation and use.

15.2 Development and Exploitation of Open Land Resources

Concepts: frontier, environmental perception

frontier

The settling of the American frontier between 1800 and 1900 is a case study of social injustice and environmental inefficiency. As population moved west, people were ruthlessly exploited to maintain slavery and enrich eastern bankers and railroad barons, while natural resources were consumed without concern for what remained.

Hardest hit during the frontier expansion were people of color: African-Americans as slaves (slavery spread westward into Kentucky, Tennessee, Missouri, and Kansas as the frontier expanded); Native-Americans were massacred and their lands stolen; Asian immigrants worked under the most difficult conditions on the expanding railroads; and entire communities of Mexican-Americans were driven from their lands in the southwest.

There was hardly more compassion for Irish immigrants who labored on the rails. And, in a cruel paradox, job opportunities in the eastern U.S. were denied to immigrants from southern Europe while the west expanded.

As people were worked to death, so was the land.

Agriculture and industry behaved like two large feet as they trampled one stock of resources after another. Wetlands were viewed as a nuisance and drained for crops. Streams were diverted for irrigation without concern for aquatic and terrestrial habitats. Entire states saw their forests disappear to provide fuel and shelter for the expanding economy. As mining gouged out the land, the poison runoff choked streams and groundwater.

environmental perception

What have been the effects of these events on our environmental perception?

There is a frontier mentality which lingers today in many MDCs which have undergone internal geographic expansion. The main feature of this mentality is the perception of unlimited resources. This belief arises from the cycle of movement characterizing frontiers—expansion and settlement are followed by more expansion and more settlement. Frontier expansion provides another example of the misuse of a common resource.

15.3 Conservation and Preservation of Open-Land Resources

Concepts: scale, exploitation, conservationism, preservationism, environmentalism, devolution, globalism

scale

Changing perspectives of environmental management often reflect the changing perceptions of scale. During different historical periods, the perception of open land and the availability of resources within those spaces have influenced our ideas about the environment (Table 15-1).

exploitation

During the period of exploitation when economic development was the primary concern, open land was viewed as unlimited. It took time, but as the land become settled, many nations finally realized the dangers of unchecked exploitation of their natural resource base. In the U.S., explorers, scientists, and intellectuals brought to public attention the values of nature. Government response reflected the two evolving approaches to natural resource use: conservationism and preservationism.

conservationism

Conservationism was based on the concepts of sustained yield and multiple use. Sustained yield was applied to forestry, and advocated harvesting timber stands at rates equal to, or less than the rate of reproduction or maturation. Multiple use encouraged recreation, grazing, mining, and irrigation activities to augment primary resource extraction or power generation within forests and watersheds. Parks were also conceived as sites for other economic uses.

preservationism

The conservation movement split during the administration of Theodore Roosevelt into the preservationist and wise-use schools. Preservationists opposed the use of forests, parks, and streams for economic uses; wise-use advocates stuck to the original conservationist goals of sustainable yield and multiple use. Both movements, though philosophically different, recognized the finiteness of open space and resources.

Modern views on the environment reflect a broadening of scale to include

TABLE 15-1 Brief History of Environmental Ideas in the United States	
Period/Ideology	Key Events and Concepts
1600-1870 / Exploitation	Homestead Act 1862, wetland drainage, clearcutting of forests, resources viewed as finite
1840-1910 / Conservationism Preservationism	Progressive reforms of Theodore Roosevelt, sustained yield, human/nature harmony, resources are finite
1930-1962 / New Deal, Government Control	New Deal reforms, Taylor Grazing Act, integrated resource management
1962-1972 / Environmentalism	Publication of Rachel Carson's Silent Spring detailing ecosystem impacts of DDT, National Environmental Policy Act (NEPA), Environmental Protection Agency formed
1976-1988 / Devolution	Clean Water Act reauthorization, control and responsibility for environmental managment shifted from federal level to states and communities
1988-present / Globalism	Montreal Protocol to limit CFCs, UNCED conference in Brazil in 1992, human environmental impacts have global consequences

concern for the global environment, while incorporating elements of the main ideologies of conservationism and preservationism. For instance, environmentalism advocates an approach which integrates the management of different resources across all media and scales. Government involvement is encouraged at all levels, particularly at the federal level. Devolution, on the other hand, advocates shifting the responsibility for environmental management from the federal to local levels of government (state and municipal). Proponents of this view claim the most efficient management of environmental problems is performed by the political institutions and agencies closest to the problem. Globalism is the recognition of human impacts at the global scale, such as those processes (deforestation, CFC use) influencing climate change.

environmentalism

devolution

globalism

15.4 Forests, Woodlands, and Land Use

Concepts: forest ecosystems, old growth forests, timber harvesting, forest fires, fuelwood

Forest ecosystems support complex interactions of plant and animal species. This

forest ecosystems

internal complexity is related to biogeography. Tropical forests have the most biodiversity, with many species of short range. Toward the higher latitudes, moisture and temperature become limiting factors, and species diversity and net productivity declines, but species range increases.

Cyclical patterns of growth and disturbance also characterize forest ecosystems. There is a typical pattern of growth in forests, similar to the "S" pattern of population growth shown in Figure 1.1. Initially the growth of young trees is slow, increases rapidly as trees mature, then slows at older ages.

Disturbances often start new cycles of succession. Pests, fire, and disease can kill all of the trees in a forest at once. Certain timber harvesting practices and shifting cultivation can also remove entire forested areas.

In the U.S., much of the debate over forest management centers around the issues of species age and timber harvesting practices. Conflicts are particularly acute over the preservation of old-growth forests. The northern spotted owl controversy in the Pacific Northwest is symbolic of the larger fight to save the old-growth forests.

old-growth forests

What is an old-growth forest?

Old-growth forests are typically older than 250 years, have big trees, big downed logs, and big standing snags. These late successional forests also have a mixed and layered canopy, a healthy understory, and light-filled gaps. Another feature is the presence of plants which fix nitrogen—ferns, moss, and lichens.

Proponents of saving old-growth forests point out that these stands store enormous amounts of carbon. There is more biomass in the temperate rain forests of the Pacific Northwest than in the tropical rain forests of Brazil. Because of this storage capability, some ecologists are suggesting the use of old-growth forests as hedges against the accumulation of the greenhouse gas carbon dioxide in the atmosphere.

Opponents of old-growth preservation cite economic necessity. Entire communities have come to depend upon timber harvesting for their sustenance. Loggers and mill operators claim the total protection of old-growth forests will result in the loss of jobs.

As frequently happens, local conflicts raise broader issues. The old-growth debate in the Pacific Northwest has called into question the basic methods used to harvest timber.

timber harvesting

Timber harvesting methods alter forest ecosystem dynamics and generate ecosystem impacts at the local, regional, and possibly global scales. Three major types of timber harvesting practices are used in the U.S.: clearcutting, selective cutting, and tree farming.

Clearcutting is the removal of all trees in a given forest stand. Selective cutting harvests only certain trees, targeting the ones with the most commercial value on the basis of size or species. Tree farming is analogous to crop farming; a single tree species is planted and harvested in a cycle, leading to stands of even-aged trees.

Table 15-2 compares the environmental impacts of the major timber harvesting practices. Concerns over timber harvesting methods are being debated for all forests, not just old-growth. With any method, human impacts are unavoidable—the goal is

TABLE 15-2 Environmental Impacts from Timber Harvesting	
Method	Impacts
Clearcutting	biodiversity decline, soil nutrient depletion, erosion, stream siltation, (FRAGMENTATION, SIMPLIFICATION)
Selective Cutting	some loss of biodiversity, erosion, removal of best trees increases susceptibility to diseases and pests (SIMPLIFICATION)
Tree Farming	monoculture represents a loss of biodiversity, even-aged stands provide poorer habitat opportunities (SIMPLIFICATION, SUBSTITUTION)

to minimize them.

forest fires

Forest ecology is also affected by fire. The key management issue is whether to fight fires or to let them burn. Again, the western United States provides an example.

Until the early 1900's, large regions of pine forest dominated the west. These forests had large trees, with understories of shrubs and grass. Periodic fires burned the understory of shrubs and grass, along with downed trees and needles. Fuel loads were low, because pine bark is thick and fire resistant. As a result, most fires caused little or no damage to the large trees; but this situation soon changed.

Forest management practices adopted in the early 20th century emphasized fire prevention and control. This created an imbalance in the natural ecosystem. Stands of small crowded trees, which normally would have been burned, were spared. This resulted in stands of larger and more crowded trees which competed for moisture and space.

The increased competition for moisture and space slowed forest growth and kept individual trees weak. In this condition, large areas of trees became less able to resist pests, drought cycles, and disease. Large areas of dead or dying trees were formed—a tinderbox waiting to explode.

If forests have not been allowed to burn often and lightly, and their excess fuels have not been removed, the nature of a fire changes when it does occur. Instead of light, cool fires, the result is a very hot and intense fire. Fires of this type are called crown fires because the entire tree burns, including the crown. The fires in Yellowstone National Park (1988), and Storm King Mountain in Colorado (1994) are examples of intense crown fires.

It will be a difficult task to restore the balance of these ecosystems. Some advocate managed burns, others the removal of dying or dead trees, and many prefer to let the natural forest ecosystem correct itself. As settlement expands to areas adjacent to and within these forested ecosystems, the problem will intensify.

Should we be patient and let nature run its course, or intervene now?

fuelwood

In other parts of the world, forests face different challenges. Billions of people rely on wood as a source of fuel, but the sustainability of the resource is threatened. According to the United Nations, many LDCs are experiencing wood shortages as land is cleared for agriculture and grazing. Over 1.2 billion people are facing an unsustainable supply of wood for the future.

15.5 Rangelands, Herding, and Ranching

Concept: common areas

Rangelands represent a diverse collection of open space ecosystems. Prairies, steppes, savannas, shrubland, open woodlands and tundra are components of this landscape. With the expansion into marginal lands, humans are converting some rangelands into urban area. The major loss of rangeland, however, is due to desertification (see section 2.6).

Ecologically, rangelands are complex. Fire, vegetation, and predator-prey interactions co-exist within a narrow window of moisture availability and soil productivity. The result is an ecosystem highly sensitive to human impacts.

For instance, the introduction of animals has altered the natural balance of rangelands. Major impacts include the compaction of soil from overgrazing, and the replacement of perennial plants with less stable annual species. Desertification is also occurring at alarming rates.

Traditional societies use rangelands for livestock herding. Sub-saharan Africa contains the greatest concentration of subsistence herders. Here, marginal land, cycles of drought, and overgrazing contribute to desertification, particularly in the Sahel.

The other major use of rangelands is for commercial ranching. This system grazes animals on large expanses of public and private land. The grasslands of South America, Australia, and the western U.S. support large herds of goats, cattle, and sheep.

common areas

Unlike other common areas, such as the oceans, rangeland common areas are often contiguous with their private counterparts--in this case ranches. In the U.S., this proximity and the low fees charged by the government for grazing induces many ranchers to use public lands. This arrangement does not provide adequate incentive to maintain the public range, and as a result, the condition of this land continues to decline.

In this policy arena, ranchers have successfully avoided attempts to modify the fee structure. This points out the need to provide consistent management for public and private lands, or both will suffer. After overgrazing renders the public lands unusable, the private lands will face increasing pressure and also decline.

15.6 The Parks Movement

Concepts: preservationism, scale/observation

Establishment of national parks preserves open space, and is an expression of preservationist ideology. Several goals characterize this movement: 1) the preservation of outstanding natural or historical areas for sightseeing and recreation; 2) the creation of wildlife refuges; 3) the need for open space to relieve overcrowding; and, 4) scientific research.

preservationism

The parks movement demonstrates the interplay of scale and observation. The movement to create ex-urban park areas was aided by the creation of urban parks. Now, city dwellers could observe the benefits of open space.

scale/observation

15.7 The Wilderness Movement

Concepts: buffers, edge effects, baseline studies

Wilderness consists of those areas undisturbed by human settlement or development. These areas are remote, difficult to access. and have a minimum size—typically 1 million acres. The majority of wilderness is within cold lands, dry lands, and mountainous regions.

The size criteria for wilderness ensures the integrity of their ecosystems by providing buffers, or shields against edge effects. Buffers are sometimes ecotones—complex transitional ecosystems existing between other ecosystems.

buffers

Edge effects are the processes occurring at the edges of ecosystems. Increased exposure to wind, light, and animal populations characterize the edges of ecosystems, and these create new habitat relationships. Figure 15.1 demonstrates the edge effects occurring in forests.

edge effects

Internal wilderness areas shielded from edge effects afford scientists the ability to make baseline studies of ecosystem relationships. Baseline studies are performed on undisturbed areas, and allow scientist to establish the natural sequence of events within ecosystems. Given this information, the ability to predict human impacts on ecosystems improves.

baseline studies

15.8 Nongovernmental Organizations That Conserve Open-Land Resources

Concept: adaptive reuse

In many nations, even the lands designated for protection are experiencing the impacts of population growth. Parks and refuges are becoming crowded, thus defeating their purpose. Smog is a problem in some U.S. national parks, and extensive use of the state parks system is threatening unique ecosystems within these areas.

Enter the railroads. Their decline offers new hope for recreation in the United States by sparking a movement for the adaptive reuse of rail corridors (Vig. 15.1).

adaptive reuse

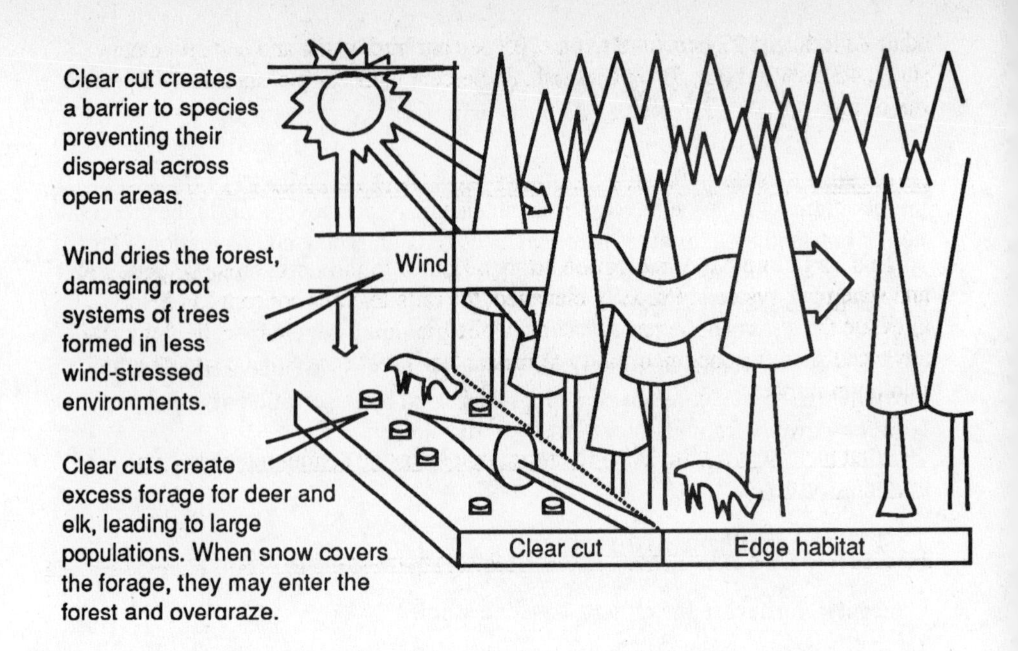

Figure 15.1 Edge effects in forests Copyright © 1995, Oregonian Publishing Co.

Vignette 15.1 Adaptive Reuse of Railroad Corridors

Some railroad history:

In 1920, there were 260,000 miles of track in rail service. By 1985, the amount of track had dropped to 145,000 miles, with analysts predicting a further decline to 100,000 miles by the year 2000. When lines are dropped from service, the adjacent land, called corridors, rapidly deteriorates. Some corridors are plowed under for agriculture, but many become dumping areas for trash.

The quantity and quality of land within rail corridors is high. There is an average of 12 acres per mile, at widths up to 400 feet. And, since railroads require gentle grades, the land is located primarily along river corridors, through mountain valleys, and across coastal plains—areas with good views and productive habitat. Many of the rail corridors also serve as habitat corridors.

Unlike railroads in other parts of the world, where railroads were superimposed on existing transportation networks, American railroads preceded development in many areas. Original ecosystems still exist in many rail rights-of-way.

People noticed the recreational and ecological potential of the abandoned rail corridors. An adaptive reuse movement called rails-to-trails arose which advocated the conversion of abandoned rail corridors to areas for hiking, biking, walking, skiing, horseback riding, and nature interpretation. These lands also provide access routes to public camping, fishing, and hunting sites. Can it work?

The demand is there. Recent studies indicate high levels of outdoor activity by adult Americans: 27 percent day hike, 35 percent bird watch and perform nature study, 46 percent bike, 51 percent fish, 84 percent walk for pleasure, and 42 percent run or jog.

This surge in outdoor recreation has contributed to the strain on the U.S. national and state park systems. Properly managed, the rails-to-trails concept can help alleviate the stress on the park systems, while providing people in technologically advanced societies the opportunity to reestablish direct ecosystem contact with minimum impacts.

What measures can be taken to ensure that set-aside common resources do not become overused?

15.9 Summary: Preservation Efforts for Earth's Ecosystems

Concept: sustainability

Open space has core areas and edges. Within the ecosystems of range, forest, and woodland, complex relationships sustain basic supplies of food, fuel, and fiber for the world's population. Human activity has altered these basic relationships, often with disastrous results.

sustainability

Sustainability of these areas requires two interrelated types of balance. The first is ecological, the second ideological. Ecological balance can be achieved through the preservation of the ecosystems, and by reducing clearcutting, overgrazing, and recreational overuse. This requires a balance of the preservationist and conservationist ideological approaches to resource management.

The historical progression from the exploitative to globalist approaches of environmental management offers hope for attaining this balance.

Applications / Discussion

15.2) Humans are part of ecosystems; therefore, when there is environmental exploitation, there must be human exploitation. Do you agree or disagree with this statement? Cite examples to support your argument.

15.3) Global forums provide the opportunity to observe a wide diversity of environmental management philosophies. Obtain a copy of the proceedings from an international forum, such as the 1992 United Nations Conference on Environment and Development (UNCED). Identify and explain the different philosophies (preservationism, conservationism, environmentalism, and globalism) present.

15.4) How does the debate surrounding the old-growth forests in the Pacific Northwest reflect the preservationism and conservationism trends in environmental ideology?

15.5) Would it be better policy to raise grazing fees so ranchers have to pay equivalent rates on public lands, or to limit the number of animals allowed on the public lands to their carrying capacity, keeping the fee constant? How could each approach fail?

15.6) How should sustainable use of a national or regional park be determined? Develop a set of criteria.

15.7) Using Figure 15.1 as a guide, visit a nearby forest and observe and record the edge effects. Make two lists: 1) the edge effects resulting from human activity; and, 2) the edge effects resulting from the transition to another ecosystem. Which source of edge effects predominates?

15.8) What are some of the social obstacles to open space set-asides, such as the rails-to-trails concept? Can you identify some of the potential environmental problems with these types of land set-aside?

Recommended Reading

Dyson, F. (1990). *From Eros to Gaia*. Pantheon, NY.

Leopold, A. (1966). *A Sand County Almanac*. Oxford Univ. Press, NY.

Lewis M., and W. Clark. (1893). *The History of the Lewis and Clark Expedition* (Vol. 1-3). Elliot Coues (ed.). Dover, NY.

McPhee, J. (1981). *Basin and Range*. Farrar, Straus and Giroux, NY.

Powell, J. W. (1961). *The Exploration of the Colorado River and its Canyons*. Dover, NY.

Sources of Additional Information

Forest and Conservation History (a periodical). Published by the Forest History Society in association with Duke University Press, 701 Vickers Ave. Durham, NC 27701

INTERNET: University of Arizona, Office of Arid Lands
e-mail: barbarah@ccit.arizona.edu

National Parks (a periodical). Published bimonthly by the National Parks and Conservation Association, 1776 Massachusetts Ave. NW Suite 200, Washington, D.C. 20036

References

Booth, W. (1989). "New Thinking on Old Growth," *Science* (244), 141-143.

Burwell, D. (1986). "Rails-to-Trails," *Wilderness*, (50), 60.

Fedkin, J. (1989). *The Evolving Use and Management of the Nation's Forests, Grasslands, Croplands, and Related Resources*. USDA Forest Service, General Technical Report RM-175. Fort Collins, CO.

Haines, L. (1986). "Rangeland Management: Whose Home is the Range?," *American Forests* (92), 22-24.

Hendee, J.C., G.H. Stankey, and R.C. Lucas. (1990). *Wilderness Management* (2nd ed.). North American Press, Golden, CO.

16

MANAGING THE ENVIRONMENT

Review

16.1 Summary

The complexity and scope of human impacts on the environment have increased rapidly since the industrial revolution. Each sphere of the geosphere, and the matter and energy linkages between them have been altered by human activity.

At the global scale, the problem of human impacts has become so complex, it defies any single classification scheme or framework. In broad overview it is useful to apply several general frameworks for evaluating human impacts. Besides giving a sense of the problem's magnitude, using several frameworks helps us consider other aspects of human impacts on the environment, and avoids tunnel vision in the problem solving process.

The misuse of technology has generated greater pressures on ecosystems, while its creative application offers powerful tools for assessing environmental impacts. Remote sensing and Geographic Information Systems are two examples of analytical tools for environmental management offering environmental scientists and planners the ability to observe, analyze, and simulate processes in the environment.

16.2 The Growing Environmental Impact of Human Activity

Concepts: four spheres, cycle of degradation, non-linear impacts, environmental equity

four spheres

Frameworks for assessing environmental impacts have already been applied at other scales and within the context of other problems, e.g., the four sphere approach. The four sphere framework allows us to gain a general perspective of the types of degradation up to a particular point in time (Table 16-1). It is useful for identifying cross-media impacts, such as acid precipitation, but is somewhat static, because it does not give us a way to estimate the future consequences of these continuing impacts.

From a temporal perspective, our use of resources has been short-sighted, favoring short term gains during a generation, rather than providing for long term sustainability. We have created a contradiction between time, matter, and energy by not using infinite resources and abusing those which are finite. Time is no longer the room for human development; at the current rates of resource use, our time is limited.

Three additional characteristics of human environmental impacts can also be identified: 1) impacts are cyclical; 2) simultaneous and successive impacts are not additive—

TABLE 16-1 Major Human Environmental Impacts on Earth within the Four Sphere Framework	
Sphere	**Major Impacts**
Lithosphere	soil erosion, desertification
Biosphere	ecosystem disruption via fragmentation, substitution, simplification, reduction, contamination, and overgrowth
Hydrosphere	point and nonpoint water pollution, dam construction and stream modifications, irrigation, groundwater overdraft
Atmosphere	air pollution from fossil fuels, acid precipitation, concentration of pollutants indoors

they are nonlinear; and, 3) the impacts are often inequitably created and distributed.

These three aspects of human environmental impacts are shown in Figure 16.1.

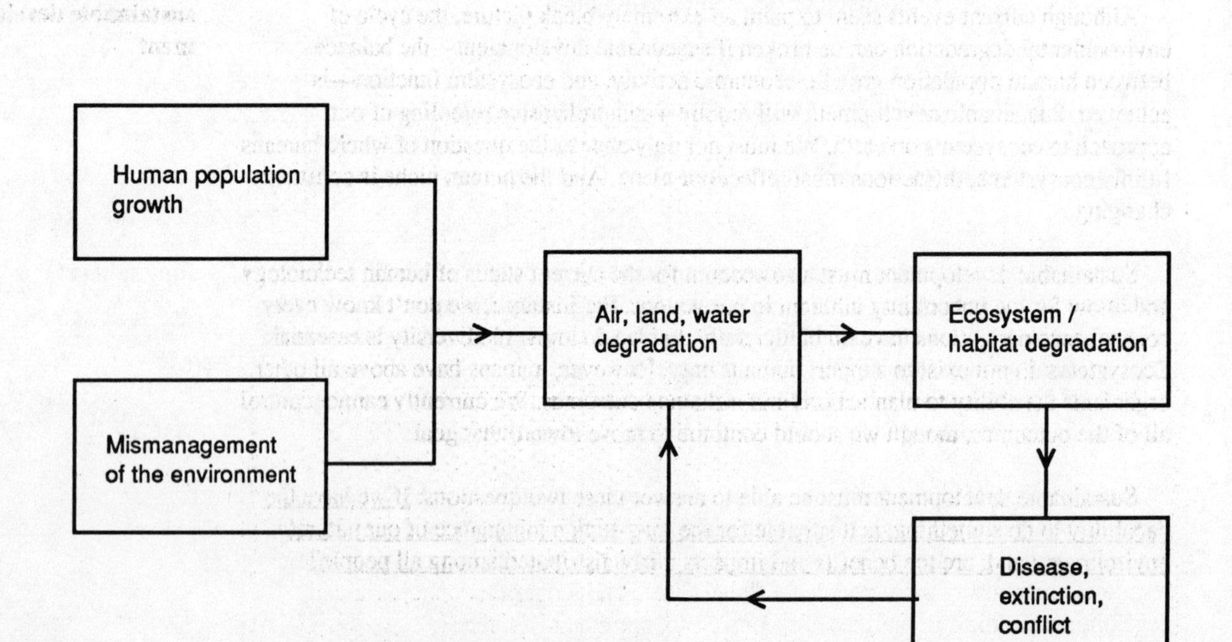

Figure 16.1 Cycle of environmental impacts

Human population growth and mismanagement of the environment have created a cycle of ecosystem degradation and social conflict. As long as the external inputs are poorly managed, the internal cycle of degradation will continue.

cycle of degradation

The nonlinear nature of environmental impacts compounds these problems. Nonlinear means the changes brought about by human activity are not strictly additive—there is an element of uncertainty governing all human environmental impacts. For example, during the recent doubling (1930 to 1975) of the world's population from 2 to 4 billion, the amount and intensity of environmental impacts more than doubled. If population doubles by the year 2035, what will be the corresponding changes in erosion, air, and water pollution?

non-linear impacts

In complex systems there is a synergism to events. Sometimes the quantity of change is less than, equal to, or more than the quantity input. For many human activities of known scale or quantity, we often don't know the resulting scale or quantity of the impacts on the environment. Finally, we must consider the character of human environmental impacts on humans themselves.

These impacts exist between and within nations, and are not borne equitably. Between nations, MDCs use 75 percent of the world's resources, but constitute only 25 percent of the world's population. Advanced technology and high energy consumption rates within MDCs deplete the resource base faster and generate more pollution per person than in the LDCs. Within all nations, poorer people and national minorities often occupy areas of poor land with high hazard potential, or live in areas susceptible to downstream or downwind water and air pollution.

environmental equity

16.3 Sustainable Development: The Key to Environmental Management

Concept: sustainable development

Although current events seem to paint an extremely bleak picture, the cycle of environmental degradation can be broken if sustainable development—the balance between human population growth, economic activity, and ecosystem function—is achieved. Sustainable development will require a comprehensive retooling of our approach to ecosystems on earth. We must not only answer the question of where humans fit into ecosystems, the actions must reflect our niche. And the human niche is constantly changing.

sustainable development

Sustainable development must also account for the current status of human technology and allow for the uncertainty inherent in our actions. For instance, we don't know every consequence our actions have on biodiversity, but we do know biodiversity is essential. Ecosystems do not exist to support humans only. However, humans have above all other organisms the ability to plan actions and influence outcomes. We currently cannot control all of the outcomes, though we should continue to move toward that goal.

Sustainable development must be able to answer these two questions: If we have the capability to do something, is it suitable for the long-term maintenance of our natural environment; and, are the benefits and impacts fairly distributed among all people?

16.4 Necessary Changes in National and International Resource Policies

Concept: global ecology

We know sustainable practices are more efficient than our current way of doing things. How do we implement them at the national and global scales? One equity-based implementation takes this form: as sustainable practices are implemented in MDCs, the savings are distributed to LDCs.

global ecology

Too idealist? OK, let's refine this concept some more. Suppose there are incentives and penalties which apply to entire nations for using sustainable or unsustainable practices in agriculture and energy use. A global ecology fund is created, with the deposits coming from fines levied against nations for unsustainable practices. To be fair, the fines should be based on ability to pay. The payment scale can be indexed by per capita Gross Domestic Product, or some other method which creates a progressive effect.

Incentives should co-exist with the penalties, or there is the possibility nations will not pursue sustainable research and development. Since MDCs are very wasteful in their energy use, and they also use the most, sustainable practices implemented here will have the widest impacts.

International negotiations would be required to set the groundrules for what constituted sustainable efforts. For instance, research into perpetual energy sources might count toward sustainable use, whereas increased militarization might count toward unsustainable practices.

Another approach would be to have all nations submit a plan for sustainable conversion of their agriculture and energy practices. Plans would have timetables for achieving objectives based upon each nation's technical and financial capacities. An international body, such as the United Nations, would approve the final plans. Failure to meet stated goals would result in a financial penalty; goal completion is rewarded.

Do you have other ideas?

16.5 Institutions to Manage the Environment

Concept: scale

scale

At the global scale, over three-quarters of the earth is public domain water—a commons. Within many nations, a large percentage of the total area exists as public land and water. The atmosphere is a public resource everywhere on earth. Who should manage these areas, and how should they be managed? How should privately owned areas be managed?

The United Nations (UN) is taking a lead role in managing the global environment. This is appropriate, since the institutional management of environmental issues should occur at a scale which accommodates the scale of the physical processes operating within ecosystems. For example, global warming is believed to result from increased emissions of carbon dioxide by fossil fuel and biomass combustion in most nations. This scale of impact required an international forum to develop a treaty to limit greenhouse gas emissions, as was done at the United Nations Conference on Environment and Development (UNCED) held in 1992.

There are also many facets of "how" public and private resources should be managed. As a start, there should be similar pricing structures for their use. Too often, private users of public resources within nations are subsidized, which contributes to their abuse. The water subsidies granted to large farms in the western U.S. (section 6.8) are an example.

16.6 Monitoring Environmental Change with Remote Sensing

Concepts: monitoring, planning, remote sensing

monitoring

Monitoring consists of measures: 1) used to evaluate work performed; or, 2) to keep track of changing conditions such as the atmospheric monitoring performed for weather reports, or periodic health examinations. Monitoring is an integral part of the planning process (Figure 16.2).

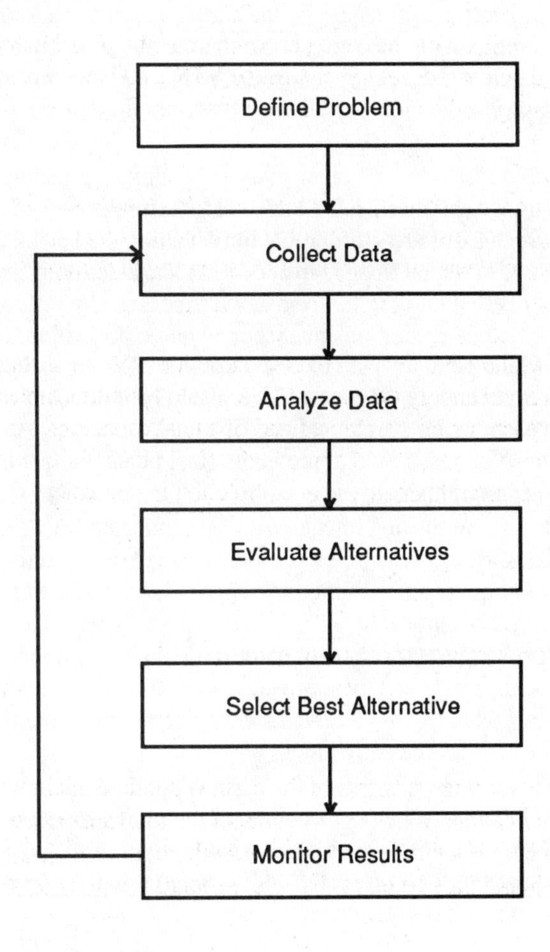

Figure 16.2 General planning process

planning

Planning is a cyclical process. Monitoring helps to ensure plan completion and the attainment of specified goals.

remote sensing

Remote sensing is a powerful monitoring tool usable during every step in the planning process. Remote sensing is the art and science of obtaining information from an object or area by a sensor not in contact with the object or area under study. The sensors record the electromagnetic energy reflected by objects, which produces different wavelength patterns, or energy signatures, according to temperature or material composition.

Cameras, line scanners, multi-spectral scanners, and radar are the main types of remote sensing instruments. Further definition and examples of their use will be discussed within the context of the planning process.

With respect to problem definition, suppose a large percentage of crops were wilting within a large area. Photographs from a camera mounted in the belly of an airplane could distinguish healthy crops from diseased crops based on the different energy signatures emitted from plants with different moisture levels. Most of the reflectance from leaves can be recorded by black and white infrared film, which is sensitive to wavelengths between 0.6um and 0.9um.

Data collection is a basic operation of remote sensing. A line scanner mounted in an airplane can detect temperature differences on the earth's surface. Thermal sensors are also used in weather forecasting to monitor surface temperatures. Dark images indicate areas emitting areas of lower temperature, and areas of higher temperature show up as brighter portions of the image.

Analysis of remotely sensed data involves many procedures, including "change mapping" and digital image enhancement. Change mapping compares images of the same area taken over time. Such maps are useful tools for detecting the spread of water pollution or crop disease. In digital image enhancement, the electronic signals produced by line scanners are converted into digital format and processed by the computer. Processing techniques include resolution enhancement and colorization. Multi-spectral scanners, such as the Multispectral Scanners and Thematic Mapper in the Landsat satellites, collect data in several wavelengths bands and permit the construction of composite color images.

Evaluation and selection of alternatives in a plan can be facilitated by comparing the objects identified from remote sensing images with "ground truth." This helps to verify the accuracy of the classification scheme used for the objects in the study area. For instance, using areas of known composition on the ground, such as a wetland, the correspondence between the sampled areas and their energy signature on a remotely sensed image can be worked out. Several computer programs perform this type of classification automatically, provided there are adequate samples or "training sets" available for comparison. This technique saves investigators time in the field by avoiding the need to observe and classify the entire area.

Finally, all types of monitoring can be performed by remote sensing instruments. Satellites produce images of the earth's surface at regular intervals, and some, such as the French SPOT system, can aim their sensors. Radar is a particularly effective monitor of landscape change because it can penetrate cloud cover.

16.7 Geographic Information Systems

Concepts: geographic information systems, spatial data

geographic information systems

Geographic information systems (GIS) are specialized computer programs designed to store, analyze, display, and output spatial data. Spatial data are data which reference real-world locations using an established coordinate system, such as latitude and longitude.

The distinguishing characteristic of GIS is their data. To see how GIS data differ, let's look at two pieces of information—called records in computerese (Figure 16.3).

Payroll Record #1

Name	Address	City	State	Salary
Jones, Joe	100 Main	Rye	CO	25000

Payroll Record #2

Name	Address	City	State	Salary	Latitude	Longitude
Jones, Joe	100 Main	Rye	CO	25000	37.85	104.91

Figure 16.3 Spatially referenced information

spatial data

Payroll record #1 and #2 are identical, except for the latitude and longitude information added to record #2. Record #2 now is spatial data, because it is spatially referenced to a coordinate system—latitude and longitude.

Assigning spatial coordinates, or geo-referencing, allows us to display spatial data within meaningful contexts of the earth's surface. For instance, Rye, Colorado is located at 37.85 north latitude and 104.91 degrees west longitude. The four corner boundaries of Colorado are located at 37 N and 102 W (southeast), 37 N and 109 W (southwest), 41 N and 109 (northwest), and 41 N and 102 W (northeast). The coordinates of Rye fall within these boundaries (Figure 16.4).

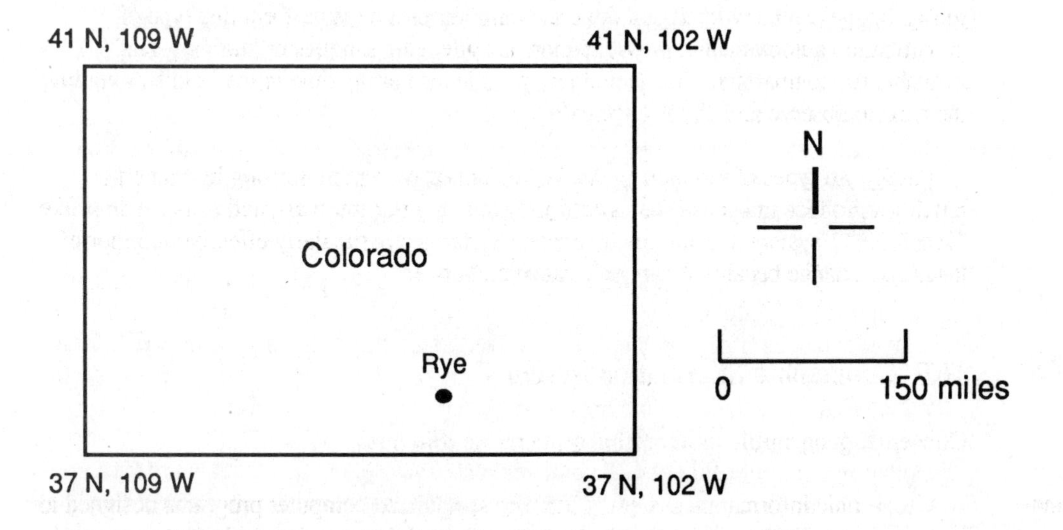

Figure 16.4 Rye, Colorado, geo-referenced

Suppose Mr. Jones had 15 coworkers at his place of employment in Rye, and they lived all over south-central Colorado. The company wished to see if a ride-sharing program would be suitable given the employee residence locations. Latitude and longitude information were added to each employee record. Now, against the backdrop of Colorado, 15 points are displayed, each representing an employee residence (Fig. 16.5).

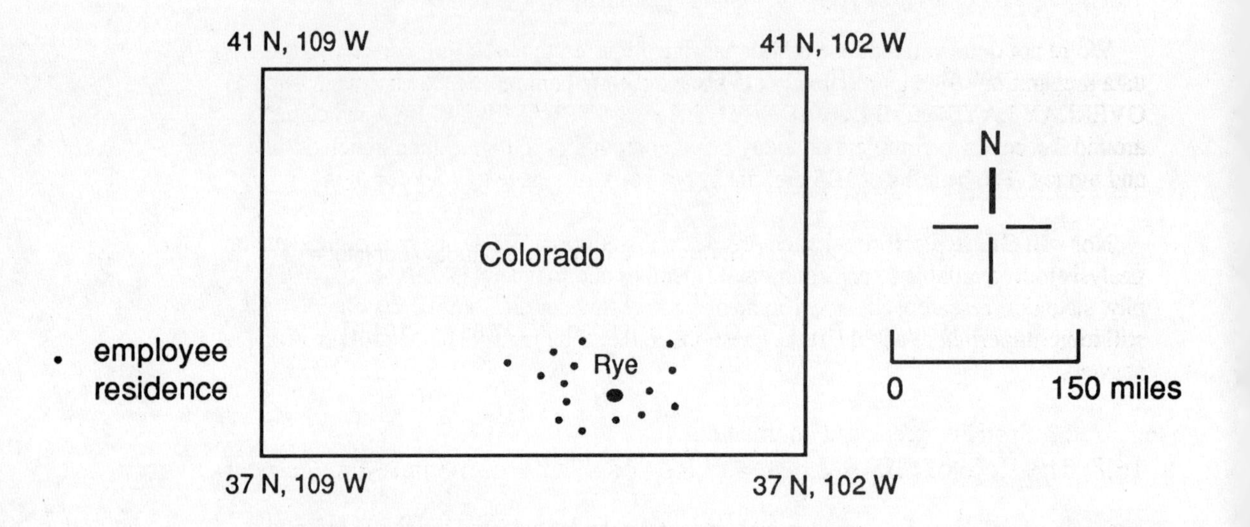

Figure 16.5 Point layer of employees

In the map above, there are two sets of data, or data layers. One layer is a base map consisting of Colorado's four corner boundary coordinates and the lines connecting these points; the second layer consists of the points representing employee residences.

In order to finish the ride-sharing analysis, we could compute the distance between each residence and develop clusters of employees in close proximity. GIS can perform both operations, which are referred to as distance and surface functions, respectively. Distance functions are operations used to compute distance, show optimal routes, or create buffer zones around map objects. Surface operations are those performed on one layer at a time.

The capability to add and overlay layers of data makes GIS a very powerful environmental analysis tool. Site suitability, zones prone to hazards, or areas impacted by human activities are some of the analyses performed with data overlays.

In the example above, we could add another layer of streams and create buffers representing their floodplains. We could then overlay this layer on the employee residence layer and determine which employees lived in the floodplain and were at risk from floods. Additional information, called attribute data, could be input for each stream, such as its flood frequency to determine how often the employees were at risk.

Another important feature of GIS data are their structure. Figures 16.16 and 16.17 in *Environmental Geography* display the two types of data structures—raster and vector. The key distinction between these data structures is geometric. A raster system uses grid cells to represent the earth's surface; vector systems use points, lines, and polygons. With geo-referenced data, both data structures can perform data overlay operations.

Data are central to GIS. Over 50 percent of the cost of a typical GIS application is related to data acquisition, preparation, and input. To conceptualize this, think about how much effort is required to locate and input each employee in the example above. Now add the time required to create the outline of Colorado and the employee location points. This requires a digitizer to create a computer file containing x,y coordinates from a paper map. This x,y coordinate file is then converted into latitude and longitude or projected into another coordinate system.

We're not done yet. Add the additional time required to check the accuracy of this data preparation. By the time you get to issue a simple command to the software, such as OVERLAY LAYER EMPLOYEE WITH LAYER STREAMBUFFERS—which takes around 2 seconds to complete on today's fast computers--you've spent a bunch of time and money. The benefits of GIS are many, but you really have to work for them.

Nor will GIS by itself provide environmental solutions. GIS is one part of a palette of analysis tools available to environmental scientists and planners. Observation, photography, statistics, research, and specific measurement devices are some of the others. It is still more important to ask the right questions, rather than to rely on one tool for your answers.

16.8 The Role of GIS in Local and Regional Environmental Management

Concepts: observation, computability

There are thousands of diverse GIS applications. For instance, GIS has been used to help direct fire trucks along the shortest routes to fires, to evaluate the spread of water pollution over time, and to assess land cover change over time. How do environmental problems become GIS applications? Let's use the response to fires as an example (Vig. 16.1).

Vignette 16.1 Fire Response—A GIS Application

observation

The process of making this GIS application begins with observation. We see a fire, and fire trucks approaching the fire via roads. The next step involves geographic synthesis, or spatial thinking—a transformation of observed phenomena into geographically related concepts. The thought process may proceed something like this: How are the locations of the fire and the location of the origin of the fire trucks related? Answer: By the roads. Then we ask: What is the location of the roads? With this line of thought, we are building spatial data relationships in our heads.

computability

The transformation into a GIS application also involves the computability of the data. This involves the determination of what layers of data are represented by the processes observed, and how these layers can be manipulated to help solve the problem. Within the context of our example, this translates into: if we want to improve the response time of fire trucks to fires, how do the relative locations of fire stations, roads, and fires interact?

One possible approach is to create a point layer of fire stations, a line layer of roads, and another point layer of fires occurring in the last 5 years within the study area. What capabilities does this information give us, and what relationships can be examined by combining these data? (Fig. 16.6).

Using the layer of fire occurrences, clusters of areas which have experienced the most fires can be identified, and a new layer showing fire-prone clusters can be created (step 1). On this new layer the location of the fire stations is overlaid (step 2). The new layer shows the proximity of the fires stations to the fire-prone areas, and helps tell us if the stations are in the best location.

We could use more of the available information. What are the shortest travel times on the roads, given the current speed limits? Now we can better determine if the stations are optimally located.

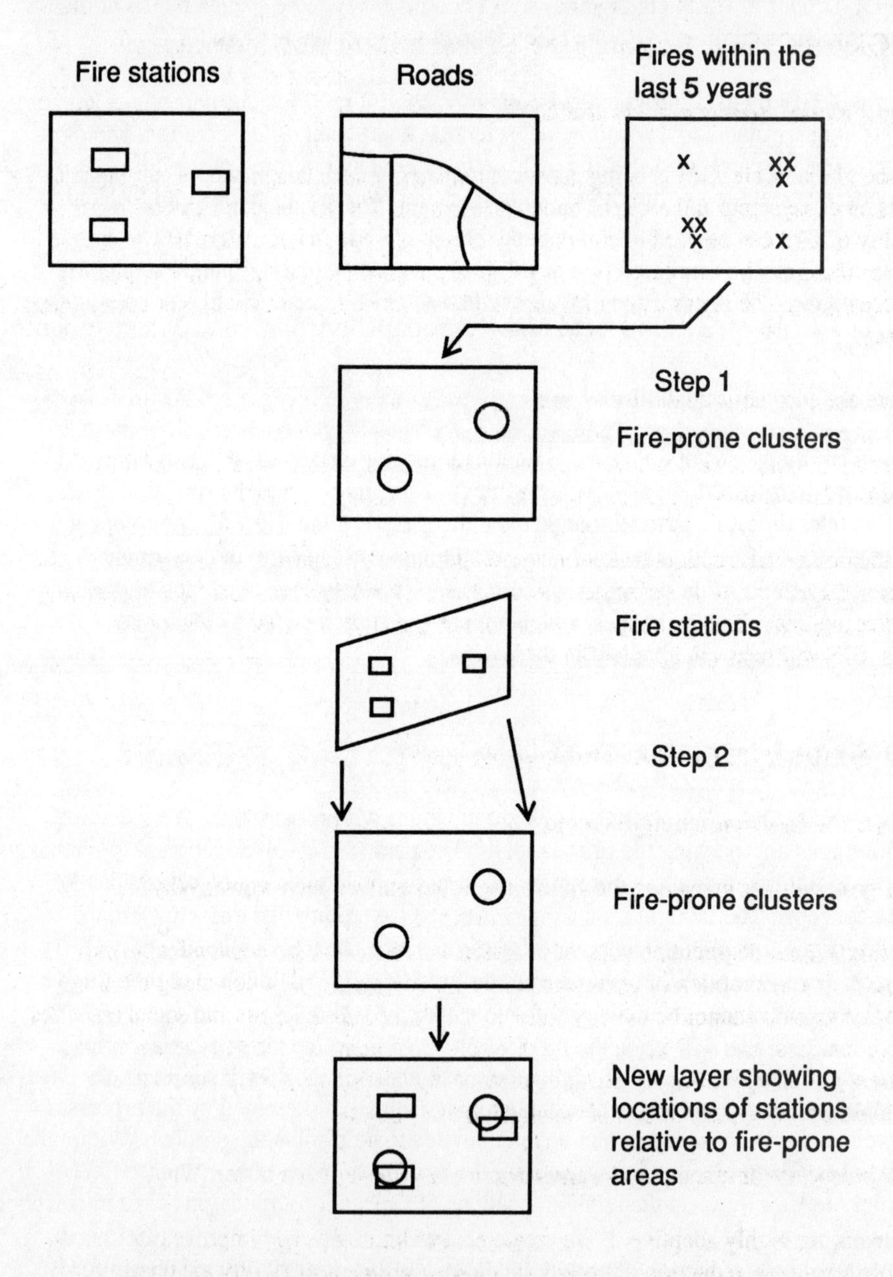

Figure 16.6 Fire response--a sample GIS application

Since all objects exist in space, there are unlimited numbers of local, regional, and global applications of GIS. The trick is to identify the problems correctly, and conceptualize the role of GIS in its solution. At every stage, if you ask the right questions, it is more likely you will get the right answers.

What types of environmental applications might not be suitable for GIS analysis?

16.9 Global Change Research and Global Information Systems

Concept: global change models and GIS

global change models and GIS

At the global scale, GIS is being used to simulate climate change and its subsequent impacts on ecosystems and areas of human settlement. For instance, the buffering capability of GIS can be used to simulate the effect of a rise in ocean levels. The spread of water inland can be compared (via overlay) to the current areas of human settlement along coastlines. The intersection of water and areas of settlement would represent zones of impact.

There are enormous quantities of data required to accurately represent the processes influencing climatic variations. There are so many "what ifs," it becomes a question of whether the critical variables have been included in the model. And, the multitude of feedback effects makes interpretation of the results difficult.

As the power (instruction execution speed and memory capacity) of computers increases, the effects of these factors will diminish. This may soon create the opportunity to predict regional climate changes, which are not currently handled by the global models. GIS will be a valuable tool in these efforts.

16.10 Summary: How Do Things Stand as We Close the Century?

Concept: the environmental dilemma

the environmental dilemma

Our prospects for managing the environment have never been worse. Why?

Environmental degradation will only worsen as LDCs develop economically and increase their consumption of resources, while emitting more pollution along the way. Population growth cannot be managed due to the traditional religious and social affinities for large families; this will strain the food supply even more in many countries, spread settlement into marginal lands, and ruin more of the landscape. And, it seems nearly impossible to obtain cooperation between nations on issues.

Our prospects for managing the environment have never been better. Why?

Humans are highly adaptive. If we create new technologies and improve our use of available resources, reduction of pollution and long-term sustainability are possible. Sustainable population growth will soon emerge if the demographic transition takes hold in LDCs. Cooperation between nations will improve if proper incentives are given for the development of sustainable land use systems, energy, and agriculture.

The optimistic views are based on "if," which means the conversion from continuing environmental disruption to global sustainability depends on the attitudes and actions of people.

It is in our hands.

Applications / Discussion

16.2) At the global scale, cite specific cases of how population growth has contributed to each of the following environmental impacts:

Air Pollution; Desertification; Energy; Waste; Erosion; Water Pollution

16.3) Sustainable development must be suitable for the long-term maintenance of our natural environment, with the benefits and impacts fairly distributed among all people. Can you improve this statement, that is, are there other considerations for sustainable development?

16.4) Outline a procedure for implementing a changeover from current energy practices to sustainable use of energy in your city. Identify the energy sources and types, and specify the general social programs or regulations needed to facilitate the conversion.

16.5) What opportunities exist at the global scale for source reduction and conservation of resources? To answer this, make a short list of 3 items currently being wasted—energy, land, and water (see example below). Next, identify the practices which waste the resource, and determine if the remedy can be achieved through source reduction, substitution, or conservation. Finally, determine the minimum institutional scale required to implement the remedy.

Resource	Wasteful Practice	Remedy	Scale
Energy	use of incandescent light bulbs	substitute fluorescents	city / local
Land	...	...	...
Water	...	...	...

16.6) Should remote sensing efforts respond to hypothesized changes in global climate, or should we define baseline environmental variables to monitor over time? Why should we do one or the other, or both?

16.7) If we are not careful, the use of GIS will be similar to giving a ray-gun to a child. What measures can prevent this from happening?

16.8) Given the following data layers, what GIS applications are possible? (assume this is for a city of 100,000 people)

Zip Code Boundaries (polygon)
Road network (all streets) (line)
Toxic sites (dry cleaners, underground tanks, landfills) (points)

Possible applications:

1. determine the concentrations of toxic sites
2. ...

16.9) Make a list of 5 "what ifs" to illustrate the complexity of global change models. For instance, what if the oceans are poorer absorbers of carbon (from carbon dioxide gas) than we thought, or, what if undiscovered microorganisms absorb methane, a greenhouse gas?

16.10) Conduct a class debate on the issue of global environmental management. The sides consist of optimists and pessimists—those who believe the prospects are good or poor for global environmental management.

Recommended Reading

Huxhold, W.E. (1991). *An Introduction to Urban Geographic Information Systems.* Oxford Univ. Press, NY.

Piel, G. (1992). *Only One World: Our Own to Make and Keep.* W.H. Freeman, NY.

Readings from Scientific American, (1990). *Managing Planet Earth.* W.H. Freeman, NY.

Sources of Additional Information

INTERNET: The Consortium on International Earth Science Information Network (CIESIN), has several digital street and boundary files available using the File Transfer Protocol (FTP). The telnet address is: ftp.ciesin.org

United Nations Environment Programme, 2 United Nations Plaza, Room DC2-303, New York, NY 10017, (212) 963-8093

United States Geological Survey, Earth Science Information Center, Denver CO 80294 (303) 844-4169

Urban and Regional Information Systems Association, 900 Second St., NE, Suite 304, Washington, D.C. 20002, (202) 289-1685

References

Burrough, P.A. (1986). *Principles of Geographical Information Systems for Land Resources Assessment.* Oxford Univ. Press, Oxford, England.

Lillesand, T.M., and R.W. Kiefer. (1994). *Remote Sensing and Image Interpretation* (3rd ed.). Wiley, NY.

Martin, James. (1976). *Principles of Data-Base Management Methodologies.* Prentice-Hall, Englewood Cliffs, NJ.

Robinson, J.L. (1988). "Mapping Mentally," *Canadian Geographer* (108), 21-23.

Tomlin, C.D. (1990). *Geographic Information Systems and Cartographic Modeling.* Prentice-Hall, Englewood Cliffs, NJ.

U.S. National Research Council, Committee on the Human Dimensions of Global Change. (1992). *Global Environmental Change: Understanding the Human Dimensions.* Academy Press, Washington, D.C.

APPENDIX - DATA APPLICATIONS

2.8 Adjacent counties to the west and south of Kansas City, MO and Kansas City, KS

County	Land Area (sq. mi.)	Population
Johnson	477	355,054
Douglas	457	81,798
Miami	577	23,466
Franklin	574	21,994
Anderson	583	14,638
Osage	704	15,248
Source: U.S. Census Bureau, 1990 Census of Population and Housing		

4.4) Country list and per capita income in current U.S. $ at market exchange rates (1990)

Country	Per capita income	Limitations of plant productivity related to climate? 0=no, 1=yes	Major biodiversity loss? 0=no, 1=yes
Africa			
Algeria	1,888		
Angola	1,131		
Botswana	2,662		
Burkina Faso	250		
Burundi	208		
Camaroon	1,160		
Cape Verde	905		
Central African Republic	487		
Chad	220		

Country	Per capita income	Limitations of plant productivity related to climate? 0=no, 1=yes	Major biodiversity loss? 0=no, 1=yes
Comoros	449		
Congo	1,274		
Djibouti	1,267		
Egypt	837		
Equatorial Guinea	463		
Ethiopia	129		
Gabon	3,823		
Gambia	346		
Ghana	415		
Guinea	480		
Ivory Coast	998		
Guinea-Bissau	243		
Kenya	368		
Lesotho	333		
Liberia	431		
Libya	8,966		
Madagascar	256		
Malawi	224		
Mali	272		
Mauritania	581		
Maritius	581		
Morocco	1,036		

Country	Per capita income	Limitations of plant productivity related to climate? 0=no, 1=yes	Major biodiversity loss? 0=no, 1=yes
Mozambique	93		
Namibia	1,474		
Niger	324		
Nigeria	299		
Reunion	8,327		
Rwanda	332		
Sao Tome	464		
Senegal	754		
Seychelles	5,254		
Sierra Leone	132		
Somalia	74		
South Africa	2,691		
Sudan	646		
Swaziland	785		
Tanzania	98		
Togo	502		
Tunisia	1,542		
Uganda	178		
Zaire	98		
Zambia	524		
Zimbabwe	676		

Country	Per capita income	Limitations of plant productivity related to climate? 0=no, 1=yes	Major biodiversity loss? 0=no, 1=yes
North America			
Anguilla	58		
Antigua/Barbuda	412		
Bahamas	11,929		
Belize	2,079		
Bermuda	27,098		
British Virgin Islands	9,566		
Canada	21,273		
Cayman Islands	18,210		
Costa Rica	1,874		
Cuba	1,968		
Dominica	2,323		
Domican Republic	991		
El Salvador	989		
Grenada	2,202		
Guadeloupe	7,015		
Guatemala	832		
Haiti	384		
Honduras	1,149		
Jamaica	1,715		
Martinique	9,907		
Mexico	2,889		

Country	Per capita income	Limitations of plant productivity related to climate? 0=-no, 1=yes	Major biodiversity loss? 0=no, 1=yes
Montserrat	6,061		
Netherlands Antilles	8,907		
Nicaragua	450		
Panama	2,072		
Puerto Rico	9,233		
St. Kitts/Nevis	2,999		
St. Lucia	1,932		
St. Vincent	1,817		
Trinidad and Tobago	4,139		
United States	21,861		
South America			
Argentina	4,373		
Bolivia	771		
Brazil	3,178		
Chile	2,110		
Columbia	1,247		
Ecuador	1,008		
Guyana	322		
Paraguay	1,231		
Peru	1,739		
Suriname	4,087		
Uruguay	2,677		

Country	Per capita income	Limitations of plant productivity related to climate? 0=no, 1=yes	Major biodiversity loss? 0=no, 1=yes
Venezuela	2,515		
<u>Asia</u>			
Armenia	4,364		
Azerbaijan	3,094		
Bahrain	7,759		
Bangladesh	212		
Bhutan	184		
Brunei Darussalam	13,969		
Cambodia	103		
Cyprus	7,850		
Georgia	4,615		
Hong Kong	12,583		
India	358		
Indonesia	580		
Iran	9,236		
Iraq	3,954		
Israel	11,734		
Japan	23,734		
Jordan	965		
Kazakhstan	4,161		
Korea, Democratic P.R.	997		
Korea, Republic of	5,626		

Country	Per capita income	Limitations of plant productivity related to climate? 0=no, 1=yes	Major biodiversity loss? 0=no, 1=yes
Kuwait	12,572		
Kyrgyzstan	2,870		
Laos	206		
Lebanon	573		
Malaysia	2,388		
Maldives	655		
Mongolia	1,096		
Myanmar	566		
Nepal	154		
Oman	6,913		
Pakistan	399		
Philippines	706		
Qatar	17,238		
Saudi Arabia	5,581		
Singapore	12,963		
Sri Lanka	461		
Syria	1,935		
Tajikistan	2,067		
Thailand	1,560		
Turkey	2,709		
Turkmenistan	3,111		
United Arab Emirates	21,259		

Country	Per capita income	Limitations of plant productivity related to climate? 0=no, 1=yes	Major biodiversity loss? 0=no, 1=yes
Uzbekistan	2,371		
Viet Nam	133		
Yemen	550		
<u>Europe</u>			
Albania	668		
Austria	20,509		
Belarus	5,868		
Belgium	19,294		
Bosnia/Herzegovina	2,886		
Bulgaria	2,305		
Croatia	5,358		
Czech Republic	3,050		
Denmark	25,242		
Estonia	7,566		
Finland	27,608		
France	21,020		
Germany	33,354		
Greece	6,587		
Holy See (Vatican)	18,916		
Hungary	3,132		
Iceland	23,622		
Ireland	12,312		

Country	Per capita income	Limitations of plant productivity related to climate? 0=no, 1=yes	Major biodiversity loss? 0=no, 1=yes
Italy	18,986		
Latvia	6,978		
Liechtenstein	50,714		
Lithuania	6,172		
Luxembourg	24,100		
Malta	6,544		
Moldova	4,362		
Monaco	21,017		
Netherlands	18,976		
Norway	24,853		
Poland	1,631		
Portugal	6,048		
Romania	1,621		
Russian Federation	6,341		
San Marino	18,916		
Slovakia	2,561		
Slovenia	8,860		
Spain	12,623		
Sweden	26,605		
Switzerland	33,674		
Ukraine	4,769		
United Kingdom	17,055		

Country	Per capita income	Limitations of plant productivity related to climate? 0=no, 1=yes	Major biodiversity loss? 0=no, 1=yes
Yugoslavia	3,849		
Oceania			
Australia	17,343		
Cook Islands	3,051		
Fiji	1,691		
French Polynesia	15,189		
Kiribati	514		
Marshall Islands	1,664		
Nauru	13,832		
New Caledonia	14,954		
New Zealand	12,909		
Papua New Guinea	831		
Samoa	713		
Solomon Islands	554		
Tonga	1,053		
Tuvalu	673		
Vanuatu	1,024		
Source: United Nations. (1994). Statistical Yearbook, 39th ed.			

5.7 Total fertility rates and average marraiage age of women in 3 MDCs and 3 LDCs.

	Total fertility rate	Average Marriage age of women
MDCs		
Japan	1.7	23.7
Sweden	1.9	26.4
United States	1.9	24.2
LDCs		
Algeria	5.4	19.5
Guatemala	5.8	21.2
Jordan	6.2	16.2
Source: United Nations. (1990). Demographic Yearbook		

Credits

CHAPTER 1

Figure 1.3: Tom and DeeAnn McCarthy/The Stock Market

CHAPTER 2

Figure 2.2: adapted from Curwen and Hatt (1953). Figure 2.3: adapted from original artwork by William M. Marsh. Figure 2.4: top: Bureau of Reclamation, U.S. Department of the Interior; middle: Bureau of Reclamation, U.S. Department of the Interior; bottom: courtesy the Soil Conservation Service

CHAPTER 3

Figure 3.1: from Marsh, W.M., and J. Dozier. (1981). *Landscape: An Introduction to Physical Geography*. Wiley. Figure 3.2: from Marsh, W.M., and J.M. Grossa Jr., (1995). *Environmental Geography: Science, Land Use, and Earth Systems*. Wiley.

CHAPTER 4

Figure 4.2: from Marsh, W.M., and J. Dozier. (1981). *Landscape: An Introduction to Physical Geography*. Wiley.

CHAPTER 5

Figures 5.1, 5.4: from Marsh, W.M., and J.M. Grossa Jr., (1995). *Environmental Geography: Science, Land Use, and Earth Systems*. Wiley. Figure 5.2: from De blij, H., and P. Muller. (1994).

CHAPTER 6

Figure 6.2: from Lappe, F.M. (1975). *Diet for a Small Planet*. Used by Permission.

CHAPTER 7

Figure 7.1: adapted from original artwork by William M. Marsh

CHAPTER 8

Figures 8.1, 8.5: adapted from Miller and Thompson, (1975). Figures 8.2, 8.7: from Marsh, W.M., and J.M. Grossa Jr., (1995). *Environmental Geography: Science, Land Use, and Earth Systems*. Wiley. Figure 8.4: from Marsh, W.M., and J. Dozier. (1981). *Landscape: An Introduction to Physical Geography*. Wiley.

CHAPTER 9

Figure 9.3: courtesy of the U.S. Environmental Protection Agency

CHAPTER 10

Figure 10.1: from Marsh, W.M., and J.M. Grossa Jr., (1995). *Environmental Geography: Science, Land Use, and Earth Systems*. Wiley. Figure 10.3: J. Skelton/USGS. Figures 10.4, 10.5: adapted from Viessman, W., and C. Welty. (1985).

CHAPTER 11

Figure 11.1: adapted from Stapp, W., and M. Mitchell. (1991). *Field Manual for Water Quality Monitoring* (6th ed.). Thompson-Shore.

CHAPTER 13

Figures 13.1, 13.2, 13.3, 13.4: from Marsh, W.M., and J. Dozier. (1981). *Landscape: An Introduction to Physical Geography*. Wiley.

CHAPTER 14

Figure 14.2: from Marsh, W.M., and J.M. Grossa Jr., (1995). *Environmental Geography: Science, Land Use, and Earth Systems*. Wiley. Figure 14.3: adapted from Mills, E.L., J.H. Leach, et al. (1994). "Exotic Species and the Integrity of the Great Lakes," *Bioscience* (44), 666-676.

CHAPTER 15

Figure 15.1: adapted from a drawing by Steve Cowden of the Oregonian staff.

Index

T

U

V

W

Y

Z